Globalized Curriculum Methods for Modern Mathematics Education

Adedeji Tella
University of Ibadan, Nigeria

A volume in the Advances in Educational Technologies and Instructional Design (AETID) Book Series

IGI Global
DISSEMINATOR OF KNOWLEDGE

Published in the United States of America by
IGI Global
Information Science Reference (an imprint of IGI Global)
701 E. Chocolate Avenue
Hershey PA, USA 17033
Tel: 717-533-8845
Fax: 717-533-8661
E-mail: cust@igi-global.com
Web site: http://www.igi-global.com

Library of Congress Cataloging-in-Publication Data

Names: Tella, Adedeji, 1973- editor.
Title: Globalized curriculum methods for modern mathematics education / Adedeji Tella, editor.
Description: Hershey PA : Information Science Reference, [2019] | Includes bibliographical references.
Identifiers: LCCN 2018002502| ISBN 9781522561583 (h/c) | ISBN 9781522561590 (eISBN)
Subjects: LCSH: Mathematics--Study and teaching--Nigeria. | Mathematics--Study and teaching--Rwanda. | Mathematics--Study and teaching--South Africa. | Mathematics--Study and teaching--Thailand. | Curriculum planning--Nigeria. | Curriculum planning--Rwanda. | Curriculum planning--South Africa. | Curriculum planning--Thailand.
Classification: LCC QA14.N6 G56 2019 | DDC 510.71--dc23 LC record available at https://lccn.loc.gov/2018002502

This book is published in the IGI Global book series Advances in Educational Technologies and Instructional Design (AETID) (ISSN: 2326-8905; eISSN: 2326-8913)

British Cataloguing in Publication Data
A Cataloguing in Publication record for this book is available from the British Library.

For electronic access to this publication, please contact: eresources@igi-global.com.

Advances in Educational Technologies and Instructional Design (AETID) Book Series

ISSN:2326-8905
EISSN:2326-8913

MISSION

Education has undergone, and continues to undergo, immense changes in the way it is enacted and distributed to both child and adult learners. From distance education, Massive-Open Online-Courses (MOOCs), and electronic tablets in the classroom, technology is now an integral part of the educational experience and is also affecting the way educators communicate information to students.

The **Advances in Educational Technologies & Instructional Design (AETID) Book Series** is a resource where researchers, students, administrators, and educators alike can find the most updated research and theories regarding technology's integration within education and its effect on teaching as a practice.

COVERAGE

- Classroom Response Systems
- Instructional Design Models
- Instructional Design
- Virtual School Environments
- Game-Based Learning
- Curriculum Development
- Hybrid Learning
- Higher Education Technologies
- Digital Divide in Education
- Online Media in Classrooms

The Advances in Educational Technologies and Instructional Design (AETID) Book Series (ISSN 2326-8905) is published by IGI Global, 701 E. Chocolate Avenue, Hershey, PA 17033-1240, USA, www.igi-global.com. This series is composed of titles available for purchase individually; each title is edited to be contextually exclusive from any other title within the series. For pricing and ordering information please visit http://www.igi-global.com/book-series/advances-educational-technologies-instructional-design/73678. Postmaster: Send all address changes to above address.

Titles in this Series

For a list of additional titles in this series, please visit:
https://www.igi-global.com/book-series/advances-educational-technologies-instructional-design/73678

Enhancing Social Presence in Online Learning Environments
Michael Marmon (University of North Texas, USA)
Information Science Reference • ©2018 • 317pp • H/C (ISBN: 9781522532293) • US $185.00

Digital Technologies and Instructional Design for Personalized Learning
Robert Zheng (University of Utah, USA)
Information Science Reference • ©2018 • 391pp • H/C (ISBN: 9781522539407) • US $195.00

Curriculum Development for Gifted Education Programs
Jessica Cannaday (Azusa Pacific University, USA)
Information Science Reference • ©2018 • 284pp • H/C (ISBN: 9781522530411) • US $175.00

Curriculum Internationalization and the Future of Education
Semire Dikli (Georgia Gwinnett College, USA) Brian Etheridge (Georgia Gwinnett College, USA) and Richard Rawls (Georgia Gwinnett College, USA)
Information Science Reference • ©2018 • 360pp • H/C (ISBN: 9781522527916) • US $195.00

Handbook of Research on Integrating Technology Into Contemporary Language Learning...
Bin Zou (Xi'an Jiaotong-Liverpool University, China) and Michael Thomas (University of Central Lancashire, UK)
Information Science Reference • ©2018 • 626pp • H/C (ISBN: 9781522551409) • US $265.00

Technology Management in Organizational and Societal Contexts
Andrew Borchers (Lipscomb University, USA)
Information Science Reference • ©2018 • 365pp • H/C (ISBN: 9781522552796) • US $195.00

Engaging Adolescent Students in Contemporary Classrooms Emerging Research and ...
Prathiba Nagabhushan (St. Mary MacKillop College, Australia)
Information Science Reference • ©2018 • 216pp • H/C (ISBN: 9781522551553) • US $135.00

For an entire list of titles in this series, please visit:
https://www.igi-global.com/book-series/advances-educational-technologies-instructional-design/73678

IGI Global
DISSEMINATOR OF KNOWLEDGE

701 East Chocolate Avenue, Hershey, PA 17033, USA
Tel: 717-533-8845 x100 • Fax: 717-533-8661
E-Mail: cust@igi-global.com • www.igi-global.com

Table of Contents

Section 2
Research on Mathematics Education

Section 3
Challenges to Mathematics/Science and Modern Globalized Curriculum

Section 4
Ethnomathematics and Globalized Curriculum

Detailed Table of Contents

In the global village we live today, inter-cultural competencies are imperative for mathematics teachers if their teaching is to cater for all learners in their classrooms. The author argues that intercultural competency is the key to helping such mathematics learners achieve their potential. Content analysis is done to find out what inter-cultural competencies are required of mathematics teachers to be able to handle diverse learners. Data was also collected through observation in a class that the author taught at an ITE to find out what occurs when the researcher puts in aspects of students' diversity in teaching mathematics. The findings show that inter-cultural competency is improved through teachers' self-introspection of their practices and presumptions about other cultures, including learners' diverse cultures in the curriculum, learning about non-verbal communication in different cultures, learning about the modes of learning in different cultures, studying the power of teachers in different cultures, handling conflicts in different cultures, and others.

The chapter examined the influence of globalization of mathematics contents, teaching methods, and resource materials in colleges of education in Oyo state. One hundred mathematics students—46 from Emmanuel Alayande College of Education and 54 from Federal College of Education (Special), Oy—were sampled for the study. Descriptive survey type of design was adopted for the study. Three questionnaires were used: influence of globalization on mathematics content questionnaire (IGMCQ, r= 0.87), influence of globalization on mathematics teaching methods (IGMTM r= 0.78), and influence of globalization on mathematics resource materials (IGMRM r= 0.89) scales. Three research hypotheses were tested in this study. The data were analyzed using frequency count, for descriptive statistics. Chi-square statistics was used to test the hypotheses. The results show that there is no significant association between globalization and mathematics contents, teaching methods, and resource materials.

In the Nigerian system of education, there is need for introducing modern technology of learning mathematics so as to ease the learning of mathematics and allow students to learn anywhere. In other to suit the features globalization, Edmodo, which is one of the learning management system packages, was considered in this study. In Edmodo class, normal teaching processes can be carried out such as teaching, immediate response from students, classwork and assignment, marking and grading, while both students and parents have access to the score of the students. In this study, emphasis is being placed on integration, benefits, and how to use Edmodo in teaching and learning of mathematics. The study concluded that the use of different technological packages in teaching will extend learning of mathematics from the four walls of the classroom to learning outside the classroom, which allows students to learn anywhere and at any time.

It is a known fact that the world now is a global village. Almost every aspect of human endeavor is moving with space in this era of digitization and movement with ethos of ages. The teaching of mathematics has been polarized in term of introduction of cultural diversity from other countries. Within standardized curricula, concepts and teaching are largely dissociated from the knowledge and skills a child brings into the classroom. Unless learners realize that mathematics exist in their very own world, beyond school walls and beyond a Eurocentric worldview, many of them will continue to complain about it as boring and uninteresting. Universalizing the curriculum for the sake of simplifying assessment or selling textbooks is not going to minimize the anxiety or even hatred that many students feel towards numbers. Conversely, introducing everyday mathematics into curricula will help students understand that math is something related to their culture.

Section 2
Research on Mathematics Education

Fractions are among some concepts in mathematics that pupils in Nigeria primary find difficult to learn, and teachers find difficult to teach. These problems require solutions because mathematics knowledge is germane for technological advancement and national development. Pupils need to be actively involved in constructing their knowledge and digital mathematics learning games could be helpful. Thus, this study investigated the effects of digital mathematics games on primary pupils' achievement in fractions concepts. An appropriately validated pupil fractions achievement test (PFAT) was used to collect data for the study. Data collected were analyzed using t-test. It was revealed that pupils exposed to digital mathematics games performed significantly better than those exposed to conventional method. There is significant difference between achievements of male and female pupils exposed to game-based strategy.

This study determined the effects of data-driven instructional strategy (DDIS) on pre-service teachers' lesson preparation in mathematics. The moderating effect of gender was also examined. A pretest-posttest control group quasi-experimental design was adopted. Twenty-seven College of Education pre-service teachers participated

mediation according to different cultural traditions both between different countries and within a single country. Hence, the need to discuss the challenges and prospects of globalized science curriculum in Nigeria.

Section 4
Ethnomathematics and Globalized Curriculum

The mathematical knowledge consumed in schools can and does influence culture and communities. There is a close connection between development of culture and idea of mathematics. Cultural thinking, practices, and products are mathematically intertwined. Cultural practices show mathematical thinking and operations and culture are better communicated through them. This goes on to confirm the practical fact that mathematics can be understood in cultural game, analysis of local arts, daily work procedures, and skills. Mathematical idea as a science of logical reasoning is better presented from natural/familiar base of culture of the people or else it will not be understood ultimately. Hence, development and human progress cannot be based on it. Mathematics will be useless to humanity. Thus, students need to develop abilities, such as creativity and a sound set of research habits, as they learn the required mathematics. This study focused on ethnomathematics and modern globalized curriculum.

Globalization accelerates at an extreme rate through development of technologies of communication. However, people living in poor countries' access is very difficult because of the energy and financial demands that technology requires. One consequence of globalization for science and science education is it creates even more complex societies and challenges for indigenous communities. However, increasing local achievement in science and science education is advocated by a number of researchers in order to provide opportunities for people globally. This issue of making local knowledge part of the global brings with it the challenges of politics, history, language, economics, and ethics. The effects of globalization have been far-reaching, while the living standards of the world are still highly uneven. This study focused on ethno-science and globalization.

Preface

INTRODUCTION

Globalized Curriculum Methods for Modern Mathematics Education is a book that expanded the body of Mathematics Education curriculum literature by covering a wide range of topics affecting the profession and the field at large. The book provided students, scholars and stakeholders in education especially mathematics and science related subjects an essential resource for uncovering the latest research in mathematics education and modern globalized curriculum, development and technologies.

University educators are compelled to respond to social contexts constituted and regulated by regional, national and multinational flows of ideas and information, capital, material and artefacts. Levin discussed how many of the patterns and practices of everyday life are shifting and changing at different rates in response to the powerful global processes that often appear beyond immediate local control, belief or even comprehension. Thus, the homogenization aspects of the globalization processes is worthy of note. With this in mind, the globalization of mathematics education, if it was to exist, would be evident in the extent a specific mathematics curriculum (policy and practice) was present in all school systems.

The phenomenon of globalization has been characterized as a series of inter-related flows, including knowledge, capital, jobs, people and cultures, ecological systems. It is also associated with increasingly rapid developments in information and communication technologies (ICTs) as well as transport, facilitating cultural and economic exchange. Recent years have also seen a growing awareness of the finite nature of Earth's natural resources, and the dramatic effects on people and the environment arising from what is generally accepted, by the scientific community at least, as Climate Change—reflecting but one manifestation of our complex ecological systems. Around the world, many school students of today have grown up surrounded by ICTs from birth. Where prevailing economic and political situations permit, electronic devices of various sizes and types may be found in all classrooms, with personal access for many students. All of these things will have an impact on the future of the young people currently enrolled in formal education, and their mathematics education in particular.

This book involved both regional and inter-regional university operating in an environment exposed to the processes of globalization. The book explored issues that arise as the universal and the particular interpenetrate. Globalization or transnational effects upon university education and curricula, and addressed the issues and challenges that arise for an educator working in both pre service and regular mathematics education as a result of the tensions between global and local contexts.

The volume is sectional designed and there were 10 chapters by different authors in the book. All topics are treated from scratch and nothing is assumed. A notable feature of the book is the fact that deliberate attempts have been used to reflect authors experiences and reflections in their classroom practices and the teaching and learning of mathematics. In this book, every author is responsible for his/her own text. Biography of the contributors can be found in the appendix. All chapters are accepted based on a double-blind peer reviewed editorial process.

The book is published by IGI Global (formerly Idea Group Inc.), an international academic publisher of the "Information Science Reference" (formerly Idea Group Reference), "Medical Information Science Reference," "Business Science Reference," and "Engineering Science Reference" imprints. IGI Global specializes in publishing reference books, scholarly journals, and electronic databases featuring academic research on a variety of innovative topic areas including, but not limited to, education, social science, medicine and healthcare, business and management, information science and technology, engineering, public administration, library and information science, media and communication studies, and environmental science.

UNIQUENESS OF THE BOOK

This comprehensive and timely publication aims to be an essential reference source, building on the available literature in the field of Mathematics Education from developed and developing countries while providing further research opportunities in this dynamic field. It is hoped that this text will provide the resources necessary for policy makers, curriculum and technology developers and managers in the Mathematics Education and Curriculum field during this digital era.

TARGET AUDIENCE

Policymakers, academicians, researchers, advanced-level students, Mathematics Curriculum planner and developer, and government officials will find this text useful in furthering their research exposure to pertinent topics in Mathematics Education and assisting in furthering their own research efforts in this field.

Here is the summary of the structure of the book and its twelve chapters.

MATHEMATICS AND MODERN GLOBALIZED CURRICULUM

Chapter 1: Enhancing Intercultural Mathematics Teaching Competencies – What Must Mathematics Teachers Do?

In the global village we live today, intercultural competencies are imperative for mathematics teachers if their teaching is to cater for all learners in their classrooms. This chapter presumes that teachers without specific cultural understanding of the students they teach cannot optimize their learners' mathematics learning and achievement. I argue that intercultural competency is the key to helping such mathematics learners achieve their potential. This research is both theoretical and empirical. Content analysis is done to find out what inter-cultural competencies are required of mathematics teachers to be able to handle diverse learners. Data was also collected through observation in a class that the author taught at an ITE to find out what occurs when the researcher puts in aspects of students' diversity in teaching mathematics. The findings show that inter-cultural competency is improved through teachers' self-introspection of their practices and presumptions about other cultures, including learners' diverse cultures in the curriculum, learning about non-verbal communication in different cultures, learning about the modes of learning in different cultures, studying the power of teachers in different cultures, handling conflicts in different cultures and others. Empirical class observations showed that if students' cultures are incorporated in mathematics teaching; in this case counting in students' home languages, the learning atmosphere became electric, and student-student and students-lecturer affinity was boosted. Implications for global mathematics teaching are offered.

Chapter 2: Influence of Globalization on Mathematics Content, Teaching Methods, and Resource Materials in Colleges of Education in Oyo State

The study examined the influence of globalization of Mathematics contents, teaching methods and resource materials in colleges of education in Oyo state. One hundred Mathematics students (46) from Emmanuel Alayande College of Education and (54) from Federal College of Education (Special), Oyo were sampled for the study. Descriptive survey type of design was adopted for the study. Three questionnaires were used. Influence of globalization on Mathematics content questionnaire (IGMCQ, r= 0.87), Influence of globalization on Mathematics teaching methods (IGMTM r= 0.78 and Influence of globalization on Mathematics resource materials (IGMRM r= 0.89) scales. Three research hypotheses were tested in this study. The data were analyzed using frequency count, for descriptive statistics. Chi-square statistics was used to test the hypotheses. The results show that there is no significant association between globalization and Mathematics contents, teaching methods as well as resource materials. However, if the appropriate methods of teaching are used with the modern technologies in this era of globalization in most of the Colleges of Education, this would definitely improve the attitude, interest as well as learning outcomes in Mathematics as practising teachers in future. The paper recommends that government should build a well-equipped e-library and Mathematics laboratories in all the Colleges of Education in Nigeria.

Chapter 3: Globalization of Teaching Strategies in Mathematics Education in Nigeria

In Nigerian system of education, there is need for introducing modern technology of learning mathematics so as to ease the learning of mathematics and allow students to learn anywhere. Ability to connect with other classrooms around the globe is another opportunity. In other to suit the features globalization, Edmodo which is one of the Learning management system packages was considered in this study. It ensures easy delivery of concept of learning. In Edmodo class normal teaching processes can be carried out such as teaching, immediate response from students, class work and assignment, marking and grading, while both students and parents have access to the score of the students. In this study, emphasis is being placed on integration, benefits and how to use Edmodo in teaching and learning of mathematics. The study concluded that the use of different technological packages in teaching will extend learning of mathematics from four walls of classroom to learning outside the classroom which allows students to learn anywhere and at any time. This will increase the time of study and make learning easily accessible to students. It was recommended among others, that technology based instruction should be adopted by mathematics teachers and government should build well-equipped e-library in all the public schools to improve

Chapter 4: Globalized Mathematics Curriculum – Could It Ever Be Possible?

It is a known fact that the world now is a global village. Almost every aspect of human endeavour are moving with space in this era of digitization and movement with ethos of ages. The teaching of mathematics. The teaching of mathematics has been polarized in term of introduction of cultural diversity from other countries. Within standardized curricula, concepts and teaching are largely dissociated from the knowledge and skills a child brings into the class room. Unless learners realize that mathematics exist in their very own world, beyond school walls and beyond a Eurocentric world view, many of them will continue to complain about it as boring and uninteresting. Universalizing the curriculum for the sake of simplifying assessment or selling textbooks is not going to minimize the anxiety or even hatred that many students feel towards numbers. Conversely, introducing everyday mathematics into curricula will help students understand that math is something related to their culture.

RESEARCH ON MATHEMATICS EDUCATION

Chapter 5: Digital Games for Teaching and Learning of Mathematics in Nigeria Schools

Digital games have potentials to improve teaching and learning activities. Fractions concepts are among some concepts in mathematics that pupils in Nigeria primary pupils find difficult to learn, and teachers find difficult to teach. There also exists a gap between both genders in learning mathematics. These problems require solutions because mathematics knowledge is germane for technological advancement and national development. Pupils need to be actively involved in constructing their knowledge and digital mathematics learning games could be helpful. Thus, this study investigated the effects of digital mathematics games on primary pupils' achievement in fractions concepts. An appropriately validated pupil fractions achievement test (PFAT) was used to collect data for the study and the two hypotheses raised were tested at 0.05 level of significance. Data collected were analyzed using t-test. Findings revealed that pupils exposed to digital mathematics games performed significantly better than those exposed to modified conventional method. There is significant difference between achievements of male and female pupils exposed to game based strategy. It was recommended that digital games should be adopted, and packaged for learning and teaching Mathematics. Also, digital games content and features should be gender neutral.

Chapter 6: Effects of Data-Driven Instructional Strategy on Pre-Service Teachers' Mathematics Lesson Preparation in College of Education Ikere-Ekiti

A large number of pre-service teachers in colleges of education (COE) in Ekiti state are seemingly incompetent in lesson preparation, thus affecting learning outcomes of their students. Such method like data drive instruction has not been widely used by teachers in lesson preparation, this study, therefore, investigated the effects of data-driven instructional strategy (DDIS) on pre-service teachers' lesson preparation in mathematics. Effect of gender was also examined. A pretest-posttest, control group quasi-experimental design was adopted. A sample of 27 pre-service teachers who were ready for teaching practice were trained on special teaching methods (STM) out of which 13 were randomly selected and trained on DDIS. The remaining 14 were exposed to the conventional method of teaching. The study was guided by three research questions. The data collected were analyzed using analysis of variance (ANOVA). Instruments used were: rating scale tagged Practical Teaching Assessment Scale (PTAS) (Scotts' pie = 0.87) and Data Driven Instructional Package (DDIP). There was a significant main effect of treatment on pre-service teachers' lesson preparation ($F_{(1,23)}$= 26.83, partial $\eta^2 = 0.54$.) Lesson preparations of pre-service teachers exposed to DDIS ($\overline{x} = 7.12$) were better than those of control ($\overline{x} = 4.87$). There was a significant interaction effect of treatment and teachers' gender on pre-service teachers' lesson preparation ($F_{(1,23)}$=4.37). Data-driven instructional Strategy enhanced College of Education pre-service teachers' lesson preparation in mathematics in Ekiti state. It was recommended that pre-service teachers should be encouraged to incorporate DDI strategy for teaching and learning of mathematics topics.

CHALLENGES TO MATHEMATICS SCIENCE AND MODERN GLOBALIZED CURRICULUM

Chapter 7: Prospect and Challenges of Mathematics Education in the Modern Globalized Curriculum

The pursuit of mathematics for the achievement of prosperity and advancement is very conspicuous in many national developmental plans both in developed and underdeveloped countries. In this paper, an attempt was made to examine some of the major challenges facing Mathematics Education in the Modern Globalized Curriculum and it's prospect. It was argued that the success of the teaching and learning of Mathematics could only be achieved when the subject is given its rightful place. The paper highlighted the characteristics and usefulness of Mathematics, Aims and Values of Mathematics Education in the Society. Finally, meaningful suggestions were proffered that if adopted will induce a positive change into the teaching and learning of Mathematics.

Chapter 8: The Challenges and Prospects of Globalized Science Curriculum

The processes of globalization are leading to widespread changes that are impacting on education world-wide. It has affected education profoundly and in a range of ways. However, for such hyperglobalists, there will be an increasing convergence of educational policies and practices worldwide. Global policies are mediated at the national level through differing cultural and historical traditions and thus produce different national policies in response to the same global pressures. Moreover, the implementation of such national policies in schools has the further potential for mediation according to different cultural traditions both between different countries and within a single country. Hence, the need to discuss the challenges and prospects of globalized science curriculum in Nigeria.

ETHNOMATHEMATICS AND GLOBALIZED CURRICULUM

Chapter 9: Ethnomathematics and Modern Globalized Curriculum

The mathematical knowledge consumed in schools can and does influence culture and communities. There is a close connection between development of culture and idea of mathematics. Cultural thinking, practices, and products are mathematically intertwined together. Cultural practices shows mathematical thinking and operations and culture are better communicated through them. This goes on to confirm the practical fact that mathematics can be understood in cultural game, analysis of local arts, daily work procedures and skills. Mathematical idea as a science of logical reasoning is better presented from natural/familiar base of culture of the people. Else, it will not be understood ultimately, hence development and human progress cannot be based on it. Mathematics will be useless to humanity. Thus, students needs to develop abilities, such as creativity and a sound set of research habits, as they learn the required mathematics. This study focused on ethnomathematics and modern globalized curriculum.

Chapter 10: Ethno-Science and Globalization

Globalization accelerates at an extreme rate through development of technologies of communication, however people living in poor countries access is very difficult because of the energy and financial demands that technology requires. One consequence of globalization for science and science education is it creates even more complex societies and challenges for indigenous communities. However, increasing local achievement in science and science education is advocated by a number of researchers in order to provide opportunities for people globally. This issue of making local knowledge part of the global brings with it the challenges of politics, history, language, economics and ethics. The effects of globalization have been far-reaching, while the living standards of the world are still highly uneven. This study focused on ethno-science and globalization

CONCLUSION

I was deeply honored when IGI Global invited me on Modern Globalized Curriculum (MGC) and I rose to the challenge of eliciting chapters from international scholars. This tome has become an excellent resource from 12 academicians. This *Globalized Curriculum Methods for Modern Mathematics Education* is a coherent piece of work. The approach presented here derive primarily from the contributors' backgrounds and experiences in MGC. Consequently, they write to inform Policy makers, academicians, researchers, advanced-level students, Mathematics Curriculum planner and developer, and government officials of the relationships between mathematics, sciences and the modern globalized curriculum. This book conceptualizes the challenges and the prospects of modern globalized curriculum in relation to mathematics and science subjects.

In editing this book, I attempted to select multiple distinct perspectives on MGC and its approaches. Given that multiple audiences exist for any text, I trust that the academic community will find this book a useful addition to existing literature on GMC and the need for improvements in MGC to ensure broader educational and societal development. In this place, I want to thank Jordan Tepper who supervised the book.

Adedeji Tella
University of Ibadan, Nigeria

Acknowledgment

The editor of this book volume wishes to extend his profound sincere gratitude to all those who made contributions and participated in the double-blind review process for this ground-breaking volume. I thank you all for your commitment, your keen reviews your time devoted and your always helpful comments. This book project would not have been possible without your efforts.

Lastly, my unalloyed gratitude goes to my family (wife and children), for their continued patience, understanding, support, and love throughout the period of their book project, and also to prof. J. O. Ajiboye and Prof. M. K. Akinsola for their encouragement especially when I wanted to give up.

Tella Adedeji
University of Ibadan, Nigeria

Section 1

Mathematics and Modern Globalized Curriculum

Chapter 1
Enhancing Inter-Cultural Mathematics Teaching Competencies:
What Must Mathematics Teachers Do?

Judah Paul Makonye
University of the Witwatersrand, South Africa

ABSTRACT

In the global village we live today, inter-cultural competencies are imperative for mathematics teachers if their teaching is to cater for all learners in their classrooms. The author argues that intercultural competency is the key to helping such mathematics learners achieve their potential. Content analysis is done to find out what inter-cultural competencies are required of mathematics teachers to be able to handle diverse learners. Data was also collected through observation in a class that the author taught at an ITE to find out what occurs when the researcher puts in aspects of students' diversity in teaching mathematics. The findings show that inter-cultural competency is improved through teachers' self-introspection of their practices and presumptions about other cultures, including learners' diverse cultures in the curriculum, learning about non-verbal communication in different cultures, learning about the modes of learning in different cultures, studying the power of teachers in different cultures, handling conflicts in different cultures, and others.

DOI: 10.4018/978-1-5225-6158-3.ch001

INTRODUCTION

Imagine the story given in Figure 1.

One asks; Would not a similar scenario occur in a mathematics classroom with learners from diversified backgrounds? Figure 1 illustrates the constant dilemmas that teachers encounter daily as they teach learners from multi-cultural backgrounds; a single problem begetting different answers. Yet all answers are correct depending on the cultural basis of respondents. It highlights the need for developing cultural competence for all teachers including mathematics teachers who teach mathematics across the globe.

PURPOSE AND RESEARCH QUESTION

This book chapter interrogates and discusses pre-service teacher education strategies that enhance global mathematics teaching and learning through intercultural competence. How can teacher education promote intercultural awareness and intercultural competencies in preservice mathematics teachers?

Figure 1.

Suppose you are on a boat with your mother, your spouse and your child. Suddenly, the boat begins to sink. You determine that you can only save one of the other passengers. Whom do you save?

According to survey results, 60% of Americans save their spouse, 40% save their children. The reasons typically offered run along these lines:

- Save Spouse: "My spouse is my partner for life and I can have more children.";
- Save Child: "Children represent the future, so it is vital to protect them first. Probably, my spouse would support this decision."

However, among Asian cultures, or Americans of recent Asian descent, nearly 100% of respondents state that they would save the mother. The rationale I have heard offered is this:
Save Mother: "My mother gave me life; I owe her my life. I can marry again; I can have more children, but I cannot replace my mother or otherwise repay the debt I owe her." (Texin, 2002)

***Story adapted from* Martin and Vaughn (2007)**

Many factors affect learners' achievement in mathematics at least, from the learners' side and from the teachers' side but these are still not enough. For teachers, delivery of adequate subject matter knowledge is important but might not be sufficient to inculcate learners' understanding of mathematics (Makonye, 2012). For learners, Ayotola and Adedeji (2009)'s research found that a learner's gender, age, mental ability, anxiety and mathematics self-efficacy could affect their mathematical achievement. This researcher would like to focus more on the anxiety and mathematics self-efficacy factors. These factors particularly affect learners' mathematical achievement in multi-cultural classrooms that consists of diverse learners. I argue that teachers' cultural competency in such classrooms helps to humanise the learning environment and allay minority learners' anxieties and so increases their learning mathematics self-efficacies. I ask; What are the mathematics teacher's cultural competence critical in the teaching and learning of mathematics in a multi-cultural classroom? If a teachers' cultural competence is important in helping learners learn mathematics better, to what level are current pre-service teachers or serving teachers equipped for this cultural competence that could enhance mathematical achievement of all their learners? For example, Pewewardy (1993) has argued that one of the reasons Indian children (who are a minority in the USA) encounter difficulties in schools is due to the fact that educators for a long time attempted to infuse culture into education, instead of infusing education into the culture of the Indian children. Here is a problem where educationists want the Indian children to assimilate western education versus western education assimilating Indian learners' culture at least in the schools.

It is argued that the problems of multi-cultural education is caused by discontinuities between home and school culture. This occurs as learners fail to apply ways of doing things at home productively in school set ups or fail to 'do school' (Pope, 2001) as they cannot interpret school rules using home culture lenses. The converse is equally true in that teachers also fail to interpret how these diverse learners interpret contexts that arise in school. Therefore, bridging the gap, filling the chasm between home and school culture is most important and needs to be negotiated if all learners are to learn mathematics successfully (Rosemberg, Stein & Alam, 2013). This chasm, caused by the dual and often ununderstood relationship between home and school, is seen as the basis of the educational struggles of African Americans (Pewewardy, 1993). Furthermore, Hollins and King (1994) reported that African American learners regarded schools as hostile places where they could not 'be themselves' as when they tried to do so, they are were ridiculed.

The book chapter discusses this dilemma as it relates to teaching and learning mathematics. The aim being to inform teaching strategies to inform mathematics education for sub-Saharan African countries. The chapter problematises the fact that we now live in a global village where people from different nations are becoming more and more interactive due to the pervasiveness of the internet and migration, in which national geographical boundaries have become insignificant. Now more than ever before, professionals and other workers including mathematics teachers leave their home countries to teach in other countries where their skills are in high demand. This means that everywhere, teacher training now need to prepare mathematics teachers to teach diverse learners. Across Africa, mathematics teachers migrate to teach in other African countries as in the case of South Africa which employs many foreign mathematics teachers from West Africa as well as Southern Africa. Besides content and pedagogical knowledge, what other mathematics teaching competencies do these teachers need to operate in unfamiliar environments?

The employment of mathematics teachers from foreign countries to teach in South Africa assumes that in a way mathematics is a universal subject, which though may have different curricula in different countries, share some common traits that if one is trained to teach mathematics in country A then they can be reasonable expected to also reasonably teach mathematics in another country B. Is that assumption valid? What are the limitations? What professional gaps in knowledge, skills and competences might these teachers need to fill?

Periodic international mathematics achievement comparison tests like Trends in Mathematics and Science Study (TIMSS) presume that mathematics curricula are or can be standardized. This author proposes that despite a presumably fairly uniform mathematics curriculum across the world, intercultural competence for teachers is critical to effect global mathematics teaching, to teach immigrant learners in the mathematics teachers' own countries or to teach in foreign countries should they work there.

Mansilla and Jackson (2011, p11) propose that for a citizen to have global competence, they ought to be able to investigate world problems well beyond their immediate environment and do age-appropriate research about them. They need to recognise other people's perspectives and their own and reflect on them thoughtfully and respectfully. They must be able to communicate their ideas on these issues effectively across race, religion, language and other possible barriers. Finally, they must be able to take action to improve conditions and be able to view themselves as active change agents. Thus, global competence relate to attitudes and ethics that enable us to understand the multi-versal world around us, to understand global events and take action in addressing them in our own way (Koziol et al., 2011). In this, it is important to be respectful and aim for peaceful co-existence with other people of the world; for if they are different from us, then we are different *to* them. The common understanding that every human being wants to be loved, that all experience pain, and want to be respected notwithstanding who they are, where they come from, or who they are ought to be appreciated.

The above applies to teachers as they are the most important change agents in society. Teachers are there to educate all learners in their charge irrespective of their diversity, towards valued societal knowledge and skills. To be able to do that, I argue that teachers need to acquire cultural competence so that they can successfully mediate mathematical knowledge to every of their students. Cultural competency helps in effecting quality multicultural mathematics teaching.

Across the world, there is a continuous movement of mathematics ideas and mathematics practitioners. This phenomenon is by no means new. In the middle ages, inter-continental merchants spread mathematical ideas from the Far-East through to Western Europe. For instance, because of the influential work of Al-Khwarazm of Persia, mathematical ideas such as trigonometry and the Hindu-Arabic numeration system began to penetrate Western Europe. By the end of the middle ages, the Hindu-Arabic system had replaced the abacus system of recording and operating on numbers of Western Europe because practitioners began to recognise its clear superiority over their abacus system. This movement of mathematical ideas and mathematics educators continue unabated to this day. In the 1970s, the Russia-USA space rivalry led to the growth of new mathematical applications in space travel and rocket science.

Today global mathematics education also continues. One form of this is overseas mathematics international students. These students often return to their countries after graduation taking up prominent positions in education and influence changes in mathematics curricula and its teaching based on what they have learnt in other countries. Sometimes these professionals do not return and it becomes a brain drain to their mother countries. Nonetheless they are supposed to adapt culturally to their adopted countries in order to function. Thus the ideas of cultural competence is always important.

These days there are regular mathematics education conferences such as International Commission of Mathematics Instruction (ICMI) that commissions the International Congress of Mathematical Education (ICME) held once in four years. ICME researches ways and share effective ways of teaching mathematics across the world including multicultural mathematics education. In the same way they are other international, regional and national associations in pursuit of research and spread of research in mathematics education. A case in point South African Association for Research in Science, Technology and Mathematics Education (SAARMSTE) (Southern Africa) and Association of Mathematics Educators of South Africa (AMESA) (South Africa) Thus inter-cultural mathematics teaching is an important current issue in mathematics education today more than any other time.

SIGNIFICANCE OF THE STUDY

Since there are unbroken movements of peoples across and in countries and continents, and those who emigrate often stick to their cultures different from mainstream cultures, we need culturally competent teachers who can teach well not only the local students, but also those students who come from other cultures. We need internationally competent teachers as educational parity and quality is of necessity tied to cultural pluralism. This in itself is closely related to democracy with its esteemed trait of respect to individual opinion.

It is necessary to have policies of 'enhancing an international perspectives' in teacher education. That requires teachers to learn about their own culture as well as others' culture (Dooly & Villanueva, 2006). For this, it is important that teachers at first have intercultural awareness during pre-service or in-service training. Worldwide, education is regarded as the greatest social facility to bring about change and understanding of the world, of understanding oneself and also others, as well as understanding scientific knowledge including mathematics. Teachers play a central role in that. Education has potential to help in developing international-minded citizens, to inculcate pro-activeness in a world that is divided by race, language, religion, national boundaries to name a few. What is needed is change of attitude towards international education (Dooly & Villanueva, 2006) and mathematics education. While attitudes are important, they need to be complimented by on the cultural competences in mathematics classrooms.

LITERATURE REVIEW

Hofstede, Pedersen and Hofstede (2002) define culture as 'the collective mental programming which distinguishes the members of one group or category of people from another' (1992, p.16). Similarly, intercultural competence is the ability to mediate from one's own culture with that of others (Dlaska, 2000). These ideas are important in negotiating mathematics teaching in a global context. Banks and Banks (2013) argue that multicultural education needs to embrace ethnicity, race, social class and diversity in the classroom. The overall aim is to enable all students to learn mathematics gainfully. Also it is desirable to promote tolerance in students for people who are different from them. Noting that every learner is important, this means that multiculturalism in education cannot subscribe to the strait-jacket curriculum of 'winner take all'.

In order to have effective global education, including mathematics education, what is required is reflective praxis; combining and applying theory in practice to effect international mathematics education practices. Some authors (for instance Dooly & Villanueva, 2006) think that the most practical way to train global educators must focus not only in inter-cultural teacher training but also immersion of trainee-teachers in foreign cultures in order to experience inter-cultural problems first hand and to learn how to deal with the inter-cultural problems. Such can be done through teaching practice in a foreign country. Dooly and Villanueva (2006) argue that the first stage in that process is learning intercultural communication. This is because communication styles differ from country to country. In general communication styles can be located into the spectrum of low-context to high-context (Würtz, 2005). High-context cultures use a lot of spoken words in their communication and many meanings are implied and unsaid. Also there is a polychronic treatment of time - many things can be done simultaneously. Lateness and delays are tolerable. So is informal personal borrowing of money from each other. In these cultures symptomatic to the Orient, particularly Mexico, China and Africa, relations between people are highly valued. In low-context cultures such as the USA, UK, Germany and other western countries, great significance is given to precision of the trite written word. Communication is direct, brief and explicit. Relationships between communicators are inconsequential and there is an almost religious worship of time. There is a polychronic view of time; you only do one thing at a time as opposed to time as a flexible resource in high-context cultures. Arriving late at a pre-planned meeting without pre-noted excuse is regarded as being extremely rude, or that the participant does not value the meeting. Borrowing of any money is supposed to be done at a bank. Trying to borrow money from a person is snubbed and seen as rude or strange gesture. Thus, the study of intercultural communication such as which can take place in a mathematics classroom occurs not in a vacuum but in a cultural context that teachers need to understand and respect if they are to be effective communicators.

Zeicher (1993) proposed that teacher multicultural education must include the following: use of biography, attention to attitudinal change, provision of diverse field experience, increasing cultural knowledge, teaching about the dangers of prejudice and racism, and how to deal with it in the classrooms, and curriculum research on the relationships between, language, culture and learning.

STRATEGIES FOR CULTURAL COMPETENCES

Martin and Vaughn (2007) point out that; 'cultural competence is a key factor in enabling educators to be effective with students from cultures other than their own' (p.3).

They write:

Cultural competence is having an awareness of one's own cultural identity and views about difference, and the ability to learn and build on the varying cultural and community norms of students and their families. It is the ability to understand the within-group differences that make each student unique, while celebrating the between-group variations that make our country a tapestry. This understanding informs and expands teaching practices in the culturally competent educator's classroom. (p.4)

What is required is enhancing 'the need for educators to provide an environment of respect and reciprocity of ideas' (Martin & Vaughn, 2007, p.5). This implies that (mathematics) teachers must be open and receptive to learners from different cultures. Inter-cultural competency helps to close mathematics achievement gaps between dominant and minority groups as minority groups increase engagement in learning mathematics.

METHODOLOGY

Content analysis of literature dealing with cross cultural education and mathematics teaching and learning in particular was done. In the content analysis, journal articles, book chapters and books were analysed with respect to the promotion of intercultural competence of mathematics teachers for teaching mathematics to diversified classes.

To help strengthen the data collection and the research, empirical observations on the teaching of mathematics to student teachers at an ITE in Johannesburg was done. This investigation was done over 2015 as the author taught a course called mathematical routes to five hundred students emanating from different races and social backgrounds. There were Black, Coloured, Indian, Chinese, and White students. Besides, this group of students spoke more than twenty distinct home languages which I deliberately targeted in my multicultural teaching.

Data Analysis

Enacting Intercultural Mathematics Education at an ITE in South Africa

One of the courses that the author teaches to student teachers is called Mathematical Routes. This is a course done by all students irrespective of specialization. In each year group there are about five hundred students studying that course. The lectures are done in a lecture theatre. Students come from varied cultures and backgrounds as described above. Therefore, this is a full multi-cultural class.

Noting that every society developed mathematical practices on how to count, measure, design and locate as well as associated vocabulary to communicate compelled me to ask students to discuss mathematical ideas in their homes such as counting. In order to accommodate my students' cultures when I am teaching numeration systems, I asked the students to count from one to ten in their mother language while the rest of the class listened. This activity resulted in counting from at least twenty different languages. What I noted is that these activities animated the learning atmosphere as students noted similarities and differences in how counting is done in different languages. The overall result is that students highly enjoyed the lesson, they felt valued. Also, student-student and student-lecturer rapport improved a great deal as a result of this activity.

Also discussing the various contributions to mathematics from cultures around the world (the history of mathematics) make students proud of their cultures. They came to realize the unique achievements of different cultures and made them see that mathematics is a shared human enterprise for us all human beings.

Some Content Analysis From Literature Sources on Inter-Cultural Teaching Competency

Various themes have been proposed that could help build the inter-cultural competence of mathematics teachers. According to Martin and Vaughn (2007) there are differences in which different cultures learn. For most people from the West learning about anything occurs through the media, the internet as well as written word. Yet in other cultures, learning often occurs through oral traditions and interpersonal interaction and communication. What are the implications for this in a multicultural mathematics classroom?

As for solving problems, Martin and Vaughn (2007) observed that different cultures, when given an identical problem often arrive at different solutions (see story at introduction). This is because different cultures are guided by different ontologies and epistemologies. If a mathematics teacher does not appreciate these, there could be conflict in the classroom.

Martin and Vaughn (2007) also observed that the meaning of non-verbal communication is different from one culture to another. For instance, African children will not make direct eye contact when talking to a teacher, as in African culture that is a gesture of challenging and disrespecting elders. On the other hand, most whites would regard a child who avoids direct eye contact when talking to them as a rude and disrespectful. In some cultures, children may smile not because they are pleased but because they feel embarrassed. In other cultures, such as the Far East, smiling signals shallowness and weakness of mind. An African teacher would regard a learner who smiles as happy but those who do not as unhappy. Those people who never smile are regarded as cynical and contemptuous ingratiates. Therefore, mathematics teachers understand the meaning of non-verbal communication for learners from different cultures, they are empowered to respond appropriately to different situations for their different learners.

Martin and Vaughn (2007) have also reported on the different ways of learning in different cultures. Whereas in some western cultures students often learn from each other when they engage in interactive collaborative groups, and the teacher acts as a facilitator in student-centered lessons, in other cultures, teacher-centered classrooms are the norm. The teacher is an authority figure, the director and provider of all learning experiences. In these classes, students are expected to fear to ask the teachers questions as that is regarded as demeaning the teachers' authority. Also, in other cases, some students believe that coming early or late to class has no effect on their learning. Others may feel that as highly disrespectful and disturbing to the class. While others would feel free to question the teacher, others may prefer to grapple quietly with the learning material (Irish & Scrubb, 2012).

On conflict which is an often unavoidable in life, Martin and Vaughn (2007) also report that the way in which conflict is dealt with differs from culture to culture. While in some cultures, conflict is at least tolerated, in others, open conflict is regarded as unpleasant and demeaning. So in the USA for example, face to face resolution of conflict is seen as more desirable. In other cultures, conflicts are best resolved quietly preferably through written communication or through mediation.

Another issue refers to religion and sometimes politics. The global mathematics teacher needs to trade very carefully on those issues as they are highly emotive.

DISCUSSION

Irish and Scrubb (2012) argue that 'Robust instructional strategies and culturally sensitive curricula are critical, but more important is an instructor who is sensitive and responsive to the unique differences of each student' (p.2). Therefore, competent intercultural mathematics teaching and learning must be based on high sensitivity and meta-cognition on one's practice. This must focus on a mathematics teacher's assessment and introspection of their own predispositions and assumptions about other cultures. This is a critical stage to help dislodge latent and often tightly held cultural assumptions that might be actually misconceptions and unreasonable stereotypes hampering mathematics teacher intercultural competency.

Mathematics teacher inter-cultural competence also rests on the teachers being accepted by the communities in which they work. As a consequence, it is incumbent for the teachers to build strong relationships with their students and their students' parents. This is because relationships that are built on trust are durable and beneficial to all partners. Students who feel that they belong and are accepted are driven to excel. Common human traits such as politeness, friendliness, kindness, fairness, respect and empathy are important factors in establishing sound teacher-student relationships. As it has been seen in data analysis teachers must learn how different cultures express these pleasantries.

Irish and Scrubb (2012) emphasize the importance of accommodating and respecting each learner. If learners are respected, they respect the teacher in return. Irish and Scrubb (2012) are of the opinion that competent multicultural teachers 'not only learn from but learn about their students' (p.45). One way of accommodating learners occurs through listening to them. Learning to students must be hermeneutic and non-evaluative. Teachers must prob them and lead them by example. Teachers must listen actively and carefully to their students, putting themselves in their students' places and learning to read between cultural lines. Teachers might also listen to their students' conversations with their classmates. The things they talk about and ask teachers to discuss and explain are the areas that concern them the most.

Further, racial and minority contributions must be reflected both in the mathematics curriculum intent, enactment and assessment. Cultural competence is viewed as an ethical responsibility for a welcoming environment that is needed for all students to succeed.

CONCLUSION AND RECOMMENDATIONS

I argue that to treat unequal people equally for them to be equal is a fallacy - this position can be extended to teaching mathematics to different people in the same way. Thus, in multi-cultural mathematics classes, it is important that diverse learners are treated a little different from each other so as to cater for their peculiar needs. Even in the same cultural group children have different learning styles and show multiple intelligences (Gardner, 2003).

The changing demographics of classrooms across the world mean that it is imperative for mathematics teachers to be culturally literate. Without that intercultural competency, the teachers would not optimize all learners' mathematics achievement. Global teachers must not to just learn the basic facts, they must learn specific nuances of their students' cultures (Hodgkinson, 1991) that they can confidently act upon to facilitate mathematics learning.

What is required is not the fear of foreign cultures as we move from mono to multi-cultural societies. We need to adopt a conciliatory attitude to embrace diversity that enriches us all. Teachers may carry stereotypes of how diverse learners behave and this has to be dealt with. What is required is attitudinal change which is allied with pedagogical knowledge. Since pre-service teachers will generally have little experience with multicultural classrooms, it might be more practical for them to study multicultural competency courses either at post graduate level when they will have had such experience or through in-service professional development seminars. Also practicing teachers may make immediate use of the available opportunities (Baker, 1994). Practicing teachers could be useful resource persons as they have hands on knowledge in mathematics teacher intercultural competency.

REFERENCES

Atweh, B., Clarkson, P., & Nebres, B. (2003). Mathematics education in international and global contexts. In *Second international handbook of mathematics education* (pp. 185–229). Springer Netherlands. doi:10.1007/978-94-010-0273-8_7

Ayotola, A., & Adedeji, T. (2009). The relationship between gender, age, mental ability, anxiety, mathematics self-efficacy and achievement in mathematics. *Cypriot Journal of Educational Sciences*, *4*(2), 113–124.

Banks, J. A., & Banks, P. (2013). *Multicultural Education, Multicultural Education: Characteristics and Goals, Culture, Teaching and Learning*. John Wiley & Sons.

Dlaska, A. (2000). Integrating culture and language learning in institution-wide language programmes. *Language, Culture and Curriculum*, *13*(3), 247–263. doi:10.1080/07908310008666602

Dooly, M., & Villanueva, M. (2006). Internationalisation as a key dimension to teacher Education. *European Journal of Teacher Education*, *29*(2), 223–240. doi:10.1080/02619760600617409

Gardner, H. (2003). Multiple intelligences after twenty years. American Educational Research Association, 21.

Hofstede, G. J., Pedersen, P., & Hofstede, G. (2002). *Exploring culture: Exercises, stories and synthetic cultures*. Nicholas Brealey Publishing.

Hollins, E. R., & King, J. E. (1994). *Teaching diverse populations: Formulating a knowledge base* (Vol. 1944). SUNY Press.

Irish, C., & Scrubb, M. (2013). *Five competencies for culturally competent teaching and learning*. Retrieved from http://www.facultyfocus.com/articles/teaching-and-learning/five-competencies-for-culturally-competent-teaching-and-learning

Makonye, P. J. (2012). Learner errors on calculus tasks in the NSC examinations: Towards an analytical protocol for learner perturbable concepts in introductory differentiation. *International Journal of Learning*, *18*(6), 339–357.

Mansilla, V., & Jackson, A. (2011). *Educating for global competence*. New York, NY: Asia Society.

Martin, M., & Vaughn, B. (2007). Cultural competence: The nuts and bolts of diversity and nclusion. *Strategic Diversity & Inclusion Management, 1*(1), 31–38.

Pewewardy, C., & Hammer, P. C. (2003). *Culturally Responsive Teaching for American Indian Students*. ERIC Digest.

Pope, D. C. (2001). *Doing school: How we are creating a generation of stressed out, materialistic, and miseducated students*. Yale University Press.

Rosemberg, C. R., Stein, A., & Alam, F. (2013). At home and at school: Bridging literacy for children from poor rural or marginalized urban communities. International Handbook of Research on Children's Literacy, Learning, and Culture, 67-82.

Texin, T. (2002). *Survey: Who Do You Save?* Retrieved: November 9, 2005. http://www.i18nguy.com/adventures/whotosave.html

Würtz, E. (2005). A cross-cultural analysis of websites from high-context cultures and low-context cultures. *Journal of Computer-Mediated Communication, 11*(1), 13. doi:10.1111/j.1083-6101.2006.tb00313.x

Chapter 2
Influence of Globalization on Mathematics Content, Teaching Methods, and Resource Materials in Colleges of Education in Oyo State

Gbolagade Olosunde
Federal College of Education (SP), Nigeria

ABSTRACT

The chapter examined the influence of globalization of mathematics contents, teaching methods, and resource materials in colleges of education in Oyo state. One hundred mathematics students—46 from Emmanuel Alayande College of Education and 54 from Federal College of Education (Special), Oy—were sampled for the study. Descriptive survey type of design was adopted for the study. Three questionnaires were used: influence of globalization on mathematics content questionnaire (IGMCQ, r= 0.87), influence of globalization on mathematics teaching methods (IGMTM r= 0.78), and influence of globalization on mathematics resource materials (IGMRM r= 0.89) scales. Three research hypotheses were tested in this study. The data were analyzed using frequency count, for descriptive statistics. Chi-square statistics was used to test the hypotheses. The results show that there is no significant association between globalization and mathematics contents, teaching methods, and resource materials.

DOI: 10.4018/978-1-5225-6158-3.ch002

INTRODUCTION

Globalization is a contemporary term used in various ways. To most scholars, globalization refers to the ideological and structural changes that are directed towards an understanding of the interconnection, especially trading, of people in different parts and areas of the world. The present form of globalization involves a combination of broad, cultural, economic, political, and technological forces that are changing the ground rules for human interaction on a worldwide scale (Kachur & Harrison, 1999).

Globalization is a recent phenomenon with its root in trade and commerce. It is a process of free exchange of goods and capitals across globe. The advancement of science and technology, revolution leading to the INTERNET, fastest means of transportation. have made the whole world a "global village" and globalization a new impetus. This means that globalization is a process wherein geographic distance becomes a factor of diminishing importance in the establishment and maintenance of cross border economic, political and socio-cultural relations. There is an emergence of global mass culture or a single world civilization. It has created a knowledge-based society. There is a shifting from a "real" community to a "virtual" community. The impact of globalization on Mathematics education is quite obvious. (www.geoiirj.com, 2012).

Globalisation processes have linked education to technological and economic development. For instance, global collaboration is essential for moving Mathematics education forward in this globalised world and at the same time avoiding the colonialism of the past and allowing the discipline to play its role in bridging the ever-increasing gap between countries (Atweh & Clarkson, 2005, Udofa & Udo, 2013).

Globalization has reinforced the utilitarian approach to school Mathematics and the Western bias in the prevailing Mathematics curricula, as well as helped to globalize pervasive Mathematical ideologies. In most instances, a new-found status that Mathematics is enjoying in this era of globalization is not well deserved, as school Mathematics can no longer be considered culturally, socially, politically, nor economically neutral. In particular, school Mathematics is increasingly critiqued as a cultural homogenizing force, a critical filter for status, a perpetuator of mistaken illusions of certainty, and an instrument of power (Smith, 2000). With such concerns, it is becoming more evident that Mathematics learning and education have implications for building just and democratic societies. This is the reason why the culture of globalization has equally engendered the Millennium Development Goals (MDGs), which seek to address issues of poverty, education, equality, wealth and environment and the Education For All Goals to be achieved (Obomanu & Adaramola, 2011). In view of these, globalization has deeply affected Mathematics education in schools since Mathematics is regarded as the peak for decision making (Udofa & Udo, 2013).

The teaching of traditional Mathematics itself raises the issue of how access to Mathematical knowledge is controlled. In a more traditional classroom, there is a rigid notion of truth that is reinforced as the teacher and/or textbooks are the sole source to evaluating the correctness of an answer. Usually problems presented to students are stripped of their social and practical contexts, while the language employed elevates agency from humans to Mathematical processes.

Denvir (1988) asserts that teaching problem-solving strategies may simply involve coaching and cramming, and rehearsed responses to evaluative assessment aimed at covering up weakness and encouraging students to adopt mechanistic behaviors. Scholars such as Denvir argue that even some of the most recent research-based reforms such as teaching heuristics might not be solutions but rather symptoms of problems of modern schooling.

Cabillan (2011) has asserted that the classical way to deliver lectures needs to be changed, minimizing the use of old textbooks and offering a very lively guiding programme based upon various information sources, with revised teaching notes and opening new windows to appreciate the context of students and their creativity as individuals and as a group. Cabillan further asserted that, there is a necessity of addressing the need to shift the culture of Mathematics learning to suit the features of globalization. In this case more specific contents should covers cognitive processes in hypermedia and multimedia learning, social issues in computer-supported collaborative learning, motivation and emotion in blended learning and e-learning.

In this digital age, as the world is advancing, the use of traditional learning gadgets and learning style has changed to the use of technology/electro-mechanical devices for societies to acquire the right information skills, attitudes and techniques for sustainable growth and development. In this case, learning is easily linked across countries thereby maintaining learning standard between societies to meet the challenges of globalization.

According to Wokocha (2011) and Udofa and Udo, (2013) the globalised world was made possible by the discovery of the computer and internet which has made the world virtual by transmitting information across the globe in a matter of micro-seconds.

Moreover, the era of globalization, ICT is used as the major way of providing the learning experience of teacher training. The content of this approach does not necessarily focus on ICT skill itself but rather covers a variety of ICT application. So, ICT is now used as part of teaching methods. Here, teachers are provided with examples of ICT pedagogy integration in their training process. Captured wisdom is a resource developed by the federally funded (USA) North Central Technology in Education Consortium for teachers, school administrators and extended to adult literacy educators. It uses videotapes and the CD-ROM to help teachers to see how technology can be integrated into their work (Iheonamekwu, Nwamuo & Okunamori, 2013).

The role of Mathematics educators in the era of globalization urgent need of the day to have competent and committed teachers, who internally develop learners as competent and committed. The teacher standing at the verge of new millennium has every demand for the acquisition of updated knowledge, skills and motives for effective functioning.

Teacher preparation programmes need to be built on the 'Learning to do', 'Learning to live together', and 'Learning to be' within built framework of reference in the following areas:

- Global and the local,
- Tradition and modernity,
- Competition and equality opportunity,
- Spiritual and material dimensions,
- Universal and the individual, and
- Knowledge explosion (Jadhav, 2012)

Udofa & Udo (2013) observed that in Nigeria, Mathematics does not appear to be in tune with the developments across the world because its pedagogical approach has not changed substantially over the years. Schools do not integrate new technologies into teaching and learning. Students do not learn how to use and integrate the new technologies in order to fit in wherever they find themselves. Communication infrastructure is inadequate for our students. Most learners in remote areas cannot have access to computers or get connected when they want to learn after school hours. It has been observed that learners do not appear to use their mobile phone for learning, and the use of computer and internet is not embraced in a higher rate particularly for learning mathematics.

The study also examined the impact of globalization in the teaching of Mathematics at the Colleges of Education level as philosophy of Nigerian Certificate in Education (NCE) is inspired by the desire to help students become intellectually informed in intellectually informed in Mathematical ideas notions and skills for logical reasoning, scientific enquiry and for the pursuit of techno-scientific education. The needs to produce non-graduates but well-groomed and qualified professional teachers of Mathematics for the Basic Education levels (NCCE, 2012).

In addition, the objectives of the NCE Mathematics is to solve abstract problems through the use of Mathematics skills and idea stimulate pupils interest in Mathematics by the use of appropriate teaching/learning strategies particularly as the Basic education levels, use mathematics to solve day to day problems and in line with digital age of globalization the objective of NCE Mathematics incorporated the use of computers in solving Mathematical problems (NCCE, 2012).

The present study investigated the influence of globalization on Mathematics content, teaching methods and resource materials in the Colleges of Education in Oyo State, Nigeria.

STATEMENT OF THE PROBLEM

The need for knowledge and technology-based personnel in the area of Mathematics is very necessary in this era of globalization. It has been observed that in Nigeria. Mathematics as a subject does not appear to be in tune with the development across the length and breadth of this nation because of its pedagogical approach has not changed substantially over the years.

Therefore, the study examined how globalization influences Mathematics content, teaching methods and resource materials in Colleges of Education in Oyo State, Nigeria.

Hypotheses

The following hypotheses were tested in this study.

Ho_1: There is no significant influence in the globalization on mathematics content.
Ho_2: There is no significant influence of globalization on mathematics teaching methods.
Ho_3: There is no significant influence of globalization on resource materials.

Methodology

The research design adopted for this study is descriptive survey. The study population comprised all the Mathematics major students in Colleges of Education in Oyo State. Two Colleges of Education were purposively selected for the study stratified along State and Federal Colleges of Education. Homogenous sample of 100 students were selected and participated using the students' GPA. In all, one hundred students participated in the study.

Influence of Globalization on Mathematics Content Questionnaire (IGMCQ) using Cronbach Alpha (r = 0.87) Influence of Globalization on Mathematics Teaching Methods (IGMTM) using Cronbach Alpha (r = 0.78) and Influence of Globalization on Mathematics Resource Materials (IGRM) using Cronbach Alpha (r = 0.89) were used to collect data. The three questionnaire used four rating scales take likert format of Strongly Agree, Agree, Disagree & Strongly Disagree. The three hypotheses were tested using chi-square statistics.

RESULTS

Testing of Hypotheses

Hypothesis 1

H_0: There is no significant influence between globalization and Mathematics content

Table 1. Demographic data by colleges

	Name of College	Frequency	Percent	Valid Percent	Cumulative Percent
Valid	Federal College of Education (Sp), Oyo	54	54.0	54.0	52.0
	Alayande College of Education	46	46.0	46.0	100.0
	Total	100	100.0	100.0	

Out of the total respondents of 100 respondents, 54 of the respondents are from Federal College of Education (Special), while 46 of the respondents are from Alayande College of Education.

Table 2. Demographic data by Department

	Department/Mathematics	Frequency	Percent	Valid Percent	Cumulative Percent
Valid	PES/MAT	29	29.0	29.0	29.0
	Chemistry/MAT	2	2.0	2.0	31.0
	ECO/MAT	46	46.0	46.0	77.0
	Computer/MAT	2	2.0	2.0	79.0
	SPE /MAT	4	4.0	4.0	83.0
	Physics/MAT	12	12.0	12.0	95.0
	G&C/MAT	5	5.0	5.0	100.0
	Total	100	100.0	100.0	

Note: PES – Primary Education Studies
MAT – Mathematics
SPE – Special Education
G&C – Guidance and Counselling
ECO - Economics

Twenty-nine of the respondents are in PES department, two are in Chemistry 46 are in ECO department, two are in Computer department, four are in Special Education, 12 are in Physics department while five are in G&C department.

Table 3. Demographic data by gender

		Frequency	Percent	Valid Percent	Cumulative Percent
Valid	Male	43	43.0	43.0	43.0
	Female	57	57.0	57.0	100.0
	Total	100	100.0	100.0	

Forty-three of the respondents are male while 57 are female.

Decision from the chi-square statistic of 0.254 and its respective Asymptotic Significant (2- sided) value of 0.615 the null hypothesis is therefore not rejected and it was concluded that there is no significant influence between globalization and Mathematics content.

Globalisation* Mathematics Teaching Methods

Hypothesis 2

H_0: There is no significant influence between globalization and Mathematics Teaching Methods

Table 4. Demographic data by colleges and department

	Course Combination	Frequency	Percent	Valid Percent	Cumulative Percent
Valid	MATFCE	3	3.0	3.0	3.0
	PES/MAT Alayande	19	19.0	19.0	22.0
	CHEM FCE	6	6.0	6.0	28.0
	CHEM/MAT Alayande	1	1.0	1.0	29.0
	ECO/MAT FCE	7	7.0	7.0	36.0
	ECO/MAT Alayande	37	37.0	37.0	73.0
	ECO/MAT	2	2.0	2.0	75.0
	PHY/MAT Alayande	1	1.0	1.0	76.0
	PHY FCE	2	2.0	2.0	78.0
	SPE/MAT FCE	5	5.0	5.0	83.0
	MAT/MAT FCE	12	12.0	12.0	95.0
	G&C MAT Alayande	5	5.0	5.0	100.0
	Total	100	100.0	100.0	

Note: FCE – Federal College of Education

Alayande – Emmanuel Alayande College of Education (State own)

Three of the respondents are in PES/MAT, 19 are in PES/MAT, six are in CHEM/MAT, one is in ECO/MAT, seven are in ECO/MAT Alayande, 37 are in ECO/MAT, two are in BIO/MAT, one is in PHY FCE, two are in PHY, five are in SPE/MAT, 12 are in MAT/MAT, while 5 in G&C/MAT.

*Table 5. Globalisation * mathematics content*

			Mathematics Content		Total
			Agree	Strongly Disagree	
Globalisation	Strongly Agree Agree	Count Expected Count Count Expected Count Count Expected Count	18 16.8 30 31.2 48 48.0	17 18.2 35 33.8 52 52.0	35 35.0 65 65.0 100 100.0

*Table 6. Chi-Square showing globalisation * mathematics content*

	Value	Df	Asymp.Sig. (2- sided)	Exact Sig. (2- sided)	Exact Sig. (1- sided)
Pearson Chi-Square	.254[a]	1	.615		
Continuity Correction[b]	.086	1	.769		
Likelihood Ratio	.254	1	.615		
Fishers Exact Test				.677	.384
Linear-by-Linear Association	.251	1	.616		
N of Valid Cases	100				

a. 0 cells (0.0%) have expected count less than 5. The minimum expected count is 16.80.
b. Computed only for a 2x2 table

*Table 7. Globalisation * mathematics teaching methods*

			Mathematics Content		Total
			Agree	Strongly Disagree	
Globalisation	Strongly Agree Agree	Count Expected Count Count Expected Count Count Expected Count	15 13.7 24 25.4 39 39.0	20 21.4 41 39.7 61 61.0	35 35.0 65 65.0 100 100.0

*Table 8. Chi-Square Table showing the influence of globalisation * mathematics teaching methods*

	Value	df	Asymp.Sig. (2- sided)	Exact Sig. (2- sided)	Exact Sig. (1- sided)
Pearson Chi-Square	.337[a]	1	.562		
Continuity Correction[b]	.133	1	.715		
Likelihood Ratio	.335	1	.563		
Fishers Exact Test				.668	.356
Linear-by-Linear Association	.333	1	.564		
N of Valid Cases	100				

a. 0 cells (0.0%) have expected count less than 5. The minimum expected count is 13.65.
b. Computed only for a 2x2 table

Decision from the chi-square statistic of 0.337 and its respective Asymptotic Significant (2- sided) value of 0.562 the null hypothesis is therefore not rejected and it was concluded that there is no significant influence between globalization and Mathematics Teaching Methods

Globalisation* Mathematics Resource Material

Hypothesis 3

H_0: There is no significant influence between globalization and Mathematics Resource Material

Decision from the Chi-Square Statistic of 0.029 and its respective Asymptotic Significant (2- sided) value of 0.866 the null hypothesis is therefore not rejected and it was concluded that there is no significant influence between globalization and Mathematics Resource Material.

DISCUSSION OF FINDINGS

The finding of Ho1 reveals that there is no significant influence of globalization on Mathematics content. This is in line with the findings of Udofa & Udo (2013); Rulkwoski & Rulkwoski (2008) which says that policy and curriculum may be impacted more by regional forces than global forces which mean that students are yet to feel the impact of globalization on Mathematics contents. Also, Ajobiewe (2016) observed dearth of knowledge among the use of computer in CBT examination among the College of Education students and recommended that in the present

Table 9. Globalisation and mathematics resource materials

			Mathematics Content		Total
			Agree	**Strongly Disagree**	
Globalisation	Strongly Agree Agree	Count Expected Count Count Expected Count Count Expected Count	15 15.4 29 28.6 44 44.0	20 19.6 36 36.4 56 56.0	35 35.0 65 65.0 100 100.0

Table 10. Chi-Square showing globalisation and mathematics resource materials

	Value	df	Asymp.Sig. (2- sided)	Exact Sig. (2- sided)	Exact Sig. (1- sided)
Pearson Chi-Square	.029[a]	1	.866		
Continuity Correction[b]	.000	1	1.000		
Likelihood Ratio	.029	1	.866		
Fishers Exact Test				1.000	.518
Linear-by-Linear Association	.028	1	.867		
N of Valid Cases	100				

a. 0 cells (0.0%) have expected count less than 5. The minimum expected count is 15.40.
b. Computed only for a 2x2 table

dynamic world, the use of computer tools and techniques in everyday activities and profession has shown that everybody irrespective of training or occupation must know how to use computer.

The test of Ho2 reveals that there is no significant influence of globalization on Mathematics teaching methods. This is in agreement with the findings of Udofa & Udo (2013) and Ugwu (2013) that most schools and colleges lack Mathematics teachers. The implication is that most students are taught by unqualified mathematics teachers. Most qualified teachers were not intandem with the globalization, they were not trained in the use of internet facilities and innovative method of teaching. The finding is also in agreement with the International Mathematics Study (IMS) who performed a comprehensive survey of the teaching and learning of Mathematics in the schools of 24 countries around the world. The result revealed that American students were distinctly mediocre in mathematics when compared to their peers in most other countries (Vernille, 2002). Therefore, there is no impact of globalization on mathematics teaching and learning (Vernille, 2002).

The test of Ho3 also reveals that there is no significant influence of globalization on mathematics resource materials. This is in consonance with findings of Ugwu (2013) which shows that schools lack both teaching materials and infrastructural facilities. The implication is that mathematics is presented in abstraction manner thereby making the learning of Mathematics difficult. Worst still is the lack of infrastructural facilities which implies that students will be performing at a minimum in terms of achievement in mathematics.

Also, Cabillan (2012) findings show a very low level of the use of technology in the classrooms from four departments is also in agreement with the present findings. Therefore, there is no influence of globalization on Mathematics resources. The implication is that with globalization which enhances the adoption of good modern techniques of instructional delivery through the use of mathematics hypermedia and multimedia and demonstration procedure, students can become versatile in their approach to solving Mathematics problems (Atweh & Clarkson, 2005; Cabillah, 2011; Udofa & Udo, 2013).

With the use of internet, computer, social media, e-learning, mobile phones and information technology in learning Mathematics, students' potentials of understanding the concept of Mathematics will increase. In addition, the adoption of globalization in Mathematics teaching and learning, students are likely to develop interest in the subject. This is not the case from the present findings.

CONCLUSION

Education system of many nations are influenced by the political, economic all and cultural aspects of the society. Tremendous changes noticed were as a result of globalization. Globalization has brought changes in the field of education. The study examined how globalization influences mathematics content, teaching method and resource materials. The study revealed there is no significant influence of globalization on mathematics content, teaching methods and resource material. The findings of the study shows that colleges possessed the mathematics teaching media to a little extent instructional facilities which make learning environment conducive for students are lacking and low level use of computer technology in teaching. So, if the appropriate methods of teaching and use of modern technologies-based instruction are embraced in most of the colleges of education as found by this study are improved, it will increase the attitude, interest as well as learning outcomes in mathematics which they will use as practicing teachers in future.

Recommendations

The study recommends among others the following:

1. Information Technology literacy should be compulsory for NCE Mathematics Course.
2. A well designed teacher training program is essential to meet the demand of today's teacher who want to learn how to use ICT effectively for their teaching.
3. Students should make use of their mobile technology, social media, computer and internet to search for information on mathematics.
4. Government should build-well equipped e-library in all the Colleges of Education to improve online learning

REFERENCES

Ajobiewe, T. (2016). *Computer appreciation training programme. Federal College of Education (Special).* Ibadan Binta Image Resources Nig. Ltd.

Apple, M. W. (1992). Do the standards go far enough? Power, policy and practice in mathematics education. *Journal for Research in Mathematics Education*, *23*(5), 412–431. doi:10.2307/749562

Atweh, B., & Clarkson, P. (2005). *More perspectives on the impact of globalisation on mathematics education in higher education.* Retrieved on 14th November, 2011 from www.merga.net.au

Cabillan, M. L. (2011). *Changing landscape of mathematics in the Philippine: lesson from globalization.* Retrieved 9th May 2012 from http://vlir-piuc.slu.edu.ph

Denvir, B. (1988). What are we assessing in mathematics and what are we assessing for? In D. Pimm (Ed.), *Mathematics and teacher education* (pp. 128–140). London: Hodder & Stoughton.

Jadhav, V. N. (2012). The Role of Teacher Education in the context of Globalization. *Global online Electronic International Interdisciplinary Research Journal, 1*(1).

Kachur, J., & Harrison, T. (1999). Public education, globalization, and democracy: Whither Alberta? In T. Harrison & J. Kachur (Eds.), *Contested classrooms* (pp. xiii–xxxiv). Edmonton, Canada: University of Alberta.

Obomanu, B. J., & Adaramola, M. O. (2011). Factor related to under achievement in science technology and mathematics education (STME) in secondary school in Rivers State, Nigeria. *World Journal of Education, 1*(1). Retrieved 14th November, 2011 from www.sciedu.ca

Odili, G. (2006). *Mathematics in Nigeria secondary schools: A teaching perspective* (Rev.ed.). Rex Charles & Patrick Limited.

Olaniyan, D. A., & Obadara, O. E. (2006). Globalisation and Nigerian educational System: Opportunities and challenges. *International Journal of African and African American Studies*, *5*(2). Retrieved from https://ojcs.siue.edu

Rutkwoski, L., & Rutkwoski, D. (2008). *Globalization in Education: An empirical look at globalization in education. An Example with Trends in International Mathematics and Science Study Mathematics data.* Retrieved 9th May, 2012 from www.iea.nl

Smith, D. G. (2000). The specific challenges of globalization for teaching and vice versa. *The Alberta Journal of Educational Research*, *46*(1).

Theonundku, S., Nwamuo, C., & Okinamiri, A. (2013). ICT Pedagogy Integration in teacher training. *Nigeria Journal of Professional Teachers*, *3*(1).

Udofa, A. I., & Udo, I. E. (2013). Globalisation and Its Influence on Mathematics Education in Secondary Schools in Akwa Ibom State, Nigeria. *Journal of Educational and Social Research*, *3*(6).

Ugwu, P. N. (2013). Mathematics as a tool for rebranding Nigeria: Implication for secondary Mathematic education. *Abacus*, *38*(11).

Vernille, K. (2002). *Why are mathematics students falling behind their international peers*. National Education Policy Center. Retrieved 19th June, 2015 from www.users.math.umd.edu

Wokocha, A. M. (2011). *Globalisation, education reforms and knowledge economy in a multicultural society*. First National Conference, Institute of Education, University of Port Harcourt. Retrieved from www.goeicrij.com

Chapter 3
Globalization of Teaching Strategies in Mathematics Education in Nigeria

Oye Akinoso
University of Lagos, Nigeria

ABSTRACT

In the Nigerian system of education, there is need for introducing modern technology of learning mathematics so as to ease the learning of mathematics and allow students to learn anywhere. In other to suit the features globalization, Edmodo, which is one of the learning management system packages, was considered in this study. In Edmodo class, normal teaching processes can be carried out such as teaching, immediate response from students, classwork and assignment, marking and grading, while both students and parents have access to the score of the students. In this study, emphasis is being placed on integration, benefits, and how to use Edmodo in teaching and learning of mathematics. The study concluded that the use of different technological packages in teaching will extend learning of mathematics from the four walls of the classroom to learning outside the classroom, which allows students to learn anywhere and at any time.

DOI: 10.4018/978-1-5225-6158-3.ch003

INTRODUCTION

The teaching and learning of mathematics should not be limited to classroom practice alone. Learning of mathematics required regular practice, but the students hardly give it time apart from the knowledge gained in classroom. In order to encourage students in providing adequate time of study for mathematics, the introduction of motivational way of having the subject very close to the students is necessary, so as to have free movement of information that can be accessed any time anywhere. Then, the teaching strategy adopts in teaching the subject should be the one accepted or recognized all over the world.

Globalization is the free movement of goods, services and people across the world in a seamless and integrated manner. It is a situation in which goods and services, or social and cultural influences gradually become similar in all parts of the world. It implies the opening of local and nationalistic perspectives to a broader outlook of an interconnected and interdependent world with free transfer of capital, goods and services across national frontiers. Globalization is also a process of making something generally accepted all over the world. It has unified entire world into small village with common socio-economic, political, cultural and educational system (Wokocha, 2011). The processes of globalization have linked education to technological and economic development, so technology has been the other principal driver of globalization.

Mobile learning applications are used today among students and educators all over the world. Example of mobile devices include smart phones, PDAs and tablets, they can be used to benefit students' learning in or out of the classroom. According to Udofa and Udo (2013), the discovery of computer and internet has made globalization possible, so, the nature of teaching and learning is changing from traditional learning to electronic method. Wokocha (2011) emphasizes that, mathematics plays important role in the school curriculum in every country. It is a means of access to other courses thereby helping to develop many professionals. Cabillan (2011) suggests that there is need to shift the culture of mathematics learning to suit the features globalization. To suit the features of learning, Learning Management System must be put into consideration.

A learning management system (LMS) is a software based platform that provides infrastructure, framework and tools to facilitate online learning or training. It enables the storage and delivery of learning for the purpose of training and educating. The LMS manages all administrative aspects of the educational process and provides important services to host and deliver the training content. It is a software program that helps create, manage and deliver e-learning courses. It is a computer program that can be used to deliver education courses and training programs. It helps in organizing courses by creating, changing and assigning courses to students and grading the scores. An LMS can be used for all kinds of learning activities, it is a business tool that has been embraced by enterprises and organizations big and small. According to Cavus and Alhih (2014), learning management systems have positive implications in education.

Learning Management system can be used to deliver online courses and to train participants. It helps to organized and manages online courses, participants, results and effectiveness of courses. The typical features an LMS apart from creating courses and track the progress of students are exporting results, inviting users, hand out certificates and make integrations with third parties. It helps to create beautiful an engaging courses, manage them and send to users and the users get a distraction free learning experience.

Learning Management System can help teachers create lessons and resources that engage their students and deliver personalized learning. Teachers can also automate routine administrative tasks and communicate with colleagues, district leaders, parents and other stakeholders (Remis, 2015). It is an online platform that links teacher, students and parent together by creating group for student to register. The online tools housed in the LMS can be accessed on the many devices used by teachers and students. Learning management system which is a platform that enables the storage and delivery of learning for the purpose of training and educating include, Moodle, Edmodo, Docebo, Sakai, Interactyx, Skillsoft, SumTotal system, Eduneering, Schoology among others.

Benefits of Learning Management System

The use of LMS has the following benefits over the old way of learning.

- **Consistency of Learning:** It keeps the content centralized so that all students have a single source of content instruction and question;

- It involves easy tracking and reporting in which progress of the students can be tracked easily. One can see at a glance the level of student work;
- It creates engaging course in which text, images and video in a learning management system can be combined;
- Users learn more because, learning material is more engaging and the use of learning management system keeps users interested and challenged;
- It keeps course up-to-date;
- Course content can be distributed easily;
- It saves cost;
- Technology provides new ways of learning develop skills construct knowledge integrate project-based learning;
- It is the tools for teaching and learning of mathematics which incorporates collaboration, presentation and evaluation;
- It allows the use of interactive board;
- It involves the use of MATLAB which is an interactive programming environment for developing algorithms, analyzing and visualizing data and managing projects (Nayak, 2012).

The learning management system that was considered in this study was Edmodo. The following features can be found in Edmodo where the full discussion of the concept can be written. It consists of a place for assignment, quiz, and poll where general opinion along with question and answer can be received. It is easy to discuss, engage students, assign homework, monitor progress anytime and anywhere. Students are empowered to take control of their own progress and can gain insight into areas where they need help. Students who are reluctant to speak in class can respond to teachers questions in anonymity (Nayak, 2012). Teachers can do real-time assessments of students' collective and individual comprehension of a lesson and use that information to adjust and provide personalized instruction. Edmodo is used to take classrooms to the next level some of the uses include the following:

- It can be used to create a group and send everyone in the school the group code so as to view PowerPoint, glogster, video summaries of student projects. It is a great Virtual Exhibition to showcase student work.
- A playground group can be created where students have free reign to post whatever they want with guidelines for appropriate use of course which encourage expression and creativity but reinforce the need for boundaries and appropriate behavior.

- Create a current events group and allow students to post articles and blogs that are relevant to classroom curriculum.
- **Edmodo Field Trip:** Give students an interactive experience on a field trip with Edmodo. Provide each student with a mobile device, or divide students into small groups, ensuring at least one cell phone is available in each group. As students wander through the museum or landmark (wherever your field trip takes place), post scavenger hunt questions, polls, or writing prompts through the Edmodo mobile application, students can respond, or ask questions.
- **Create a Simple Poll Each Morning:** As students walk in the door, keep Edmodo open on your classroom computer and ask them to 'check in' on the poll on their Edmodo account.
- It can be used to create data diary. The teacher creates a data tracking group and use it as a filter on calendar.
- **Host backchannel Discussions:** Students can use the mobile application for in-class backchannel discussions during read alouds or movies. Backchannel discussions could also host outside the classroom during key events to help get students thinking about what learnt from the event and how it relates to learning in class.
- **Provide Homework Support:** Provide students with extra support by being available to answer homework questions after school hours. Allow students to direct message teacher, or post questions to the group during specific hours. If there is notifications setup for messages, immediately be alerted to new questions and can respond from mobile device.
- **Check in With Class While Out:** If the teacher is planning to be out of the classroom for any period of time, use the Edmodo application to keep in contact with students. Schedule morning check-ins with the class and have students answer questions about their learning and respond to questions from students if needed.

How to Integrate Edmodo Into Classroom Teaching and Learning

Edmodo is the premier social media and learning platform for teachers and students alike. The platform's application equivalent brings portability into the equation, which allows educators, students and parents to access Edmodo anytime. Teachers can truly bring the classroom online. With the ability to give assignments, quizzes and polls to students, Edmodo users can manage the classes and consolidate all activities in one place. Users also have unlimited options in terms of sharing digital content. Teachers can embed instructional videos, create learning groups for students, and manage calendar events. Communities allow teachers to browse subjects in order to find and share information within a broad stream of content and conversations related to that topic.

The platform offers numerous organizational features for teachers, it offers an option for users to print out rosters for substitutes. Badges are another fun feature which allows educators to monitor the progress of student groups and issue awards for the achievements. Enterprising educators can even use Edmodo to connect with other classrooms around the globe. Partnering with a class online can be socially engaging and culturally rewarding for students. Edmodo is just as great for district administrators and principals. The platform can be used to coordinate and collaborate with leadership teams, feedback can be given and receive from other educators. Planning professional development seminars can be a breeze, especially with SchoolTube, Edmodo's video service.

Students also stand to benefit from the platform. Students can access extra help from peers, instructors or SchoolTube. Students are able to turn in work for teacher review and get instant feedback, including teacher annotations. Beyond being able to join groups for clubs, sports and other activities, students can learn how to interact safely and appropriately online. On Edmodo, parents get the option to log in either as their children or themselves. Parent can review the child's work. Parents can also receive updates from child's teachers regarding missed assignments and school happenings. Edmodo helps educate parents about the digital literacy skills. (www.educationalworld.com/a_tech/how-educators-can-use-edmodo.shtml, 2017)

How to Use Edmodo in Teaching

The following chats represent the work done with Edmodo in Mth 422 class. Students were asked to register with the given code, notes and assignment were given to the students, and submission date for assignment was sent to the students. The work was marked immediately and students have access to the grades. The first chat consists of the list of students in the group, followed by note sent to students on Partial differential equation through Edmodo package, while the third chat represents the questions given as assignment, assignment followed by the response of a student. It is very easy to use just as if one is chatting on socio-media like Facebook and Whathapp.

Figure 1.

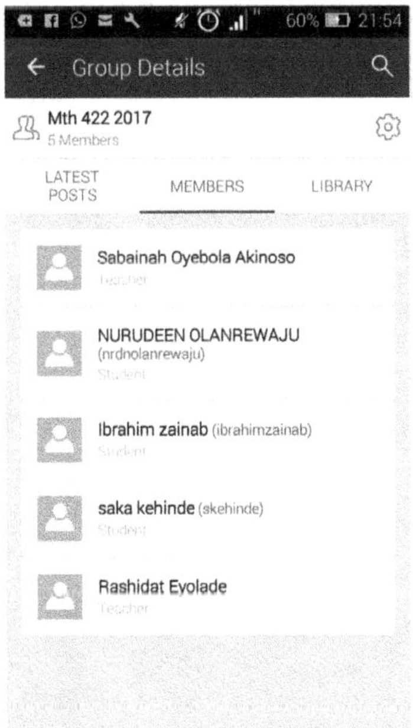

Figure 2.

Figure 3.

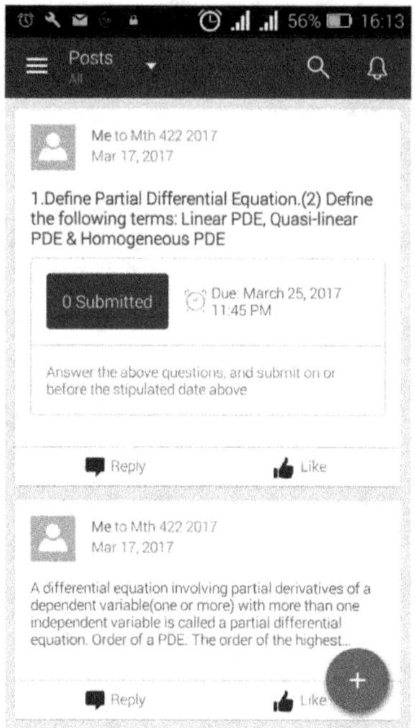

CONCLUSION

The study concluded that the use of different technological packages in teaching will extend the learning of mathematics from classroom to anywhere outside classroom which allow student to learn at any time. This will increase the time of study and make learning easily accessible to students. Edmodo can be used for concept study, mathematics projects, engaging students, host profession development session and to create group which could help improving teaching and learning of mathematics. In this case, Edmodo is a global education network that connects all learners with the people and resources needed to reach the full potential both locally and internationally. This study suggested the use of Edmodo in teaching of mathematics.

RECOMMENDATIONS

In this era of globalization, the teaching and learning of mathematics should not be based on traditional way of teaching, but the strategies recognized globally. The teacher of mathematics should be current enough to be aware of current technological system for teaching. It is recommended that the teacher of mathematics should learn the use of different Learning Management System so as to adopt them in the teaching and learning of mathematics at any level. Both teachers and students should be computer literate. Teachers should embrace computer and technology-based instructions in lessons delivery. Parents should provide computer for their children so as to interact with colleagues and teachers online and the global compliance.

REFERENCES

Cabillan, M. L. (2011). *Changing landscape of mathematics in the Philippine: Lesson from globalization.* Retrieved 1/3/2017 from http://vlir-piuc.slu.edu.ph

Carvus, N., & Alhih, M. S. (2014). Learning management systems use in science education. *Procedia: Social and Behavioral Sciences, 143*, 517–520. doi:10.1016/j.sbspro.2014.07.429

Nayak, P. (2012). *Why do a teacher use technology in teaching mathematics*. Retrieved from https://www.slideshare.net/pratimanayak39/why-do-a-tecaher-use-technology-in- teaching-mathematics Retrieved on 1/3/2017

Remis, K. K. (2015). *LMS enhances K12 instruction: Systems increase engagement, provide quick access to digital resources and help teachers with administrative tasks.* Academic Press.

Udofa A. I. & Udo I. E. (n.d.). Globalisation and its influence on mathematics education in secondary schools in Akwa Ibom state, Nigeria. *Journal of Educational and Social Research, 3*(6), 33-40.

Wokocha, A. M. (2011). Globalisation, education reforms and knowledge economy in a multicultural society. *First National Conference*, Institute of Education, University of Port Harcourt.

Chapter 4
Globalized Mathematics Curriculum:
Could It Ever Be Possible?

Adedeji Tella
University of Ibadan, Nigeria

ABSTRACT

It is a known fact that the world now is a global village. Almost every aspect of human endeavor is moving with space in this era of digitization and movement with ethos of ages. The teaching of mathematics has been polarized in term of introduction of cultural diversity from other countries. Within standardized curricula, concepts and teaching are largely dissociated from the knowledge and skills a child brings into the classroom. Unless learners realize that mathematics exist in their very own world, beyond school walls and beyond a Eurocentric worldview, many of them will continue to complain about it as boring and uninteresting. Universalizing the curriculum for the sake of simplifying assessment or selling textbooks is not going to minimize the anxiety or even hatred that many students feel towards numbers. Conversely, introducing everyday mathematics into curricula will help students understand that math is something related to their culture.

DOI: 10.4018/978-1-5225-6158-3.ch004

INTRODUCTION

It is a known fact that the world now is a global village. Almost every aspect of human endeavour are moving with space in this era of digitization and movement with ethos of ages. Everything is now globalized. The teaching enterprise is not left out most especially mathematics.

In an attempt to do justice to this paper, a reflection has to be made to my secondary school days. I remembered in the late 70s till mid 80s where expatriates from Ghana, Sri Lanka, India, Bangladesh, Ethiopia, Eritrea culminated Nigeria teaching market. These expatriates were either teaching mathematics and sciences except on few occasion where they teach Christian Religion Studies. So, in a way the teaching of mathematics has already been polarized and pluralized in term of introduction of cultural diversity from these countries on Nigerian students.

Another experience was that of brain drain whereby some scholars are leaving the shore of Nigeria for greener pasture across the global into countries such as South-Africa, USA, Australia, Belgium, Lesotho, Botswana, Kenya, Zimbabwe among others. The lesson from this was that acculturation is very imminent in such circumstance like this. There is bound to be alien introduction of culture as reflection of such in the teaching and learning to those that teach mathematics where a particular language explanation has no equivalent.

Other experience worth sharing is that of the happening in Nigeria as a result of situation whereby mathematics from the western part move to teach in the North, South and Eastern part of the country as a result of unemployment. There is no way that one culture cannot be force or introduce unknowing during the teaching and learning of mathematics. Similar to this is parent job mobility where pupils/students moved along with their parents during transfer from one location to another to earn their living. This is common among military, police, federal staff with this, cultural pollution is inevitable. This inevitability lead to multicultural education way of learning mathematics. A typical example is that of Ajegunle in Lagos state, Nigeria.

International conferences and collaborations also lend credence to the aforementioned discussions of possibility of globalized mathematics curriculum. We (mathematics educators) unknowing import foreign variables as a result of our attendance in international mathematics conferences, seminar, symposium, workshops and international collaboration with colleagues in other part of the world. We have argued elsewhere that global collaboration is essential for moving the discipline forward in this globalised world - at the same time avoiding the colonialism of the past and allowing the discipline to play its role in bridging the ever-increasing gap between countries (Atweh & Clarkson, 2001b). This issue of global collaboration sets the tone for this paper. In this study we conjecture that for those who have had and continue to have an active set of international contacts and/or experiences, their appreciation of the impact of the processes of globalization would be heightened (Clarkson & Atweh, 2004).

Another area where one imbibes these foreign cultures and pluralities are through journals reading and intellectual discourse where a foreigner is invited to discuss or share his/her experiences and findings on mathematics domain on why she/he excel in his/her teaching of mathematics regarding his/her landmarks/achievement. International assessment and comparative data in education is similarly imported, possibly after collection with local help, and is a marketable and prestigious academic product in 'developed' countries, where it features in many leading research publications. It also has extra value because it feeds into the managerialist mechanisms of education policy and control. Governments and other agencies take international comparison data as a key measure of performance and use it to evaluate and judge educational systems (analogous to the role of economic indicators). As an indicator of success in education it is a source of national pride (or shame) and it is also selectively used in the media to influence political and popular opinion.

Concept of Globalization

Educators are compelled to investigate how these thoroughgoing changes are influencing the world of learning. Perhaps surprisingly, globalization would appear in some ways to have actually caused a regression in the techniques of teaching. For the sake of streamlining the educational process, it has shifted the focus away from the priority of developing knowledge to the secondary function of assessing the results (Varghese, Thomas and McCusker, 2006). Given the seemingly limitless global pool of human resources, finding talent through mass assessment has understandably more appeal to an 'efficient' system than does the patient, innovative nurturing of its growth. And since standards are the essence of assessment, one of the consequences of globalization has been the drive to teach a universal mathematics curriculum in schools (Gates & VistroYu, 2003, Varghese, et al., 2006). The field of mathematics, perhaps because it is often thought of and presented as neutral, culturally independent and value-free, is even more susceptible to this treatment than others, and so educational systems worldwide are being prompted to mimic and mechanically copy mathematics curricula, teaching methods and assessment tests from each other (Gerdes, 1997). There is some evidence ... that the processes of globalization are compelling rather than invitational, and therefore require careful scrutiny (McGinn, 1995). And yet this social process is not necessarily negative, as some in the media would have us believe, but rather is a process that we need to be careful of. Hence this process needs to be researched is all human activity, including the cultural one of mathematics education. In passing we would comment briefly on something that Woock hints at above. Even though globalization is framed as a powerful, all encompassing process, in its turn, perhaps because of its very nature, it seems to spark or induce social processes at local levels that may be quite contrary to the globalising process, and hence effective sights of resistance come into being. Therefore, there seems to be always limits to this powerful process (Clarkson & Atweh, 2004).

Mathematics and Cultures

The conception of mathematics is a reflection of the culture and socioeconomics of any given region (Gates & Vistro-Yu, 2003), hence using a curriculum designed for one specific culture, however, might not be particularly be inspiring for learners in another culture. Compounding this problem, for most of what is generically called the 'developing world,' is the fact that the source of these 'standard' curricula is, generally speaking, the West. Therefore, globalized mathematics education not only places many Asian, African or South American students at an academic disadvantage by using curricula tailored for Western students, but also uncomfortably echoes a colonial past, the long-term effects of which are still being endured in many parts of these same regions (Varghese, et al., 2006). Dowling (1998) reminded us, the whole system will require a Westerner to reveal to non-Western students the mathematics inherent in their own, non-Western culture. When a European instructs, he or she does it in European terms, and referring to European mathematicians (Dowling cited in Varghese, et al., 2006). A prominent example of this is the present-day mathematical education in India, once part of the empire "upon which the sun never set" – an early forerunner of globalization. As in many non-Western countries using Western math curricula, it is generally taught in a narrow context with little concern for its historical and cultural settings (Varghese et al., 2006). An irrelevant and abstract curriculum, distanced from the lives of the majority, is the typical experience of school children (Rampal, 2003). The main reason for the gap between the curriculum and the lives it should enrich is that the educational system transported to India by its colonial master was not adapted to the socio-cultural context of the subcontinent.

The development of schools and institutions was conceived within the British system, not a regional one (Smith 2000; Gates & Vistro-Yu, 2003; Rampal, 2003). Even in India, where trigonometry was developed, the concept of zero was discovered and the decimal system has its roots, and which once saw mathematics as an instrument of divine intervention (Joseph, 2000), school mathematics is now more or less seen as a European product. There is evidence from the research done by Anderson (1990) that when non-Western students are taught a culturally focused mathematics curriculum, they exhibit more positive and self-assured attitudes towards mathematics, greater persistence and achievement in algebra courses, and greater rates of participation in subsequent mathematics courses when compared to their peers. Every practicing teacher is keenly aware of how important interest is to success in learning; it is so powerful, in fact, that the presence of interest in a subject almost obviates the need for discipline. It is internally rather than externally motivating (Varghese, et al., 2006). In mathematics, as in all subjects, interest is best generated by relating the concepts to everyday life, and this paper suggests certain opportunities that could allow this to happen. It is believed that mathematics education that is relevant and referent to the local socio-cultural context might work miracles of learning in rural and marginalized populations such as those of Nigeria.

Globalization and the Knowledge Economy

According to Ernest (2007), one of the defining characteristics of postmodernity is the dramatic emergence of globalization across the domains of industry, commerce, technology (including information and communication technologies), culture and education. Globalization is the social change brought about by increased connectivity among societies and their elements, including the merging and convergence of cultures. The principal means is the dramatic enhancement of transport and communication technologies used to facilitate international cultural and economic exchange. This has led to the formation of a 'global village' (McLuhan 1964), with closer contact between different parts of the world, providing increasing possibilities of personal exchange, mutual understanding and friendship between 'world citizens', which is especially notable and important in the worlds of education and research (Ernest, 2007). The outcome is the corresponding erosion of national sovereignty in the economic sphere as profit-making multinational or transnational corporations circumvent the bounds of local laws and standards through moving their operations from country to country (Ernest, 2007). Thus, globalization is often seen as global Westernization (Sen 2004).

Ethnomathematics

The idea of ethnomathematics was developed to incorporate a broader view of how mathematics relates to the real world. It is a relatively new area of mathematics education, beginning with the contributions of the Brazilian educator Ubiratan D'Ambrosio (Knijnik, 1997). The essence of ethnomathematics lies in the fact that there are distinctions between cultures. As Ascher noted (1991), differences such as where people live (deserts or islands or the Artic), how they produce food (hunting, fishing or farming) and many others affect the expression and content of their mathematical ideas. Above all, no single distinction will preclude the mathematical ideas or determine what they will be (Ascher, 1991). Mathematical activities form a significant part of peoples' everyday lives, even among unschooled communities of India (Rampal, 2003). Through cultural tools and symbolic systems, even non-literates are capable of sophisticated mental computations. As Ascher (1991) states, mathematical ideas are not "the exclusive province of some unusual few who were specially endowed at birth or specially educated" (p.188). Rampal (2003) noted that, in India, oral numeracy is more natural among unschooled adults than literacy. That means even unschooled are more familiar with numbers, operations and measurements than with letters. Therefore, there is no doubt that the children will also be familiar with the everyday mathematics their parents use. D'Ambrosio (1985) believes that children acquire the mathematical knowledge of their cultural group at a very young age. However, when they begin to attend school on a regular basis, this knowledge is replaced with school knowledge, which is more valued by the system. The inclusion of ethnomathematics (everyday mathematics) in the curriculum helps mathematics educators move in a direction that situates mathematics in real world contexts. This will be in line with the National Council of Teachers of Mathematics (NCTM) recommendations of 1989. Hence it is worth exploring how "everyday mathematics" can enrich classroom practices.

MATHEMATICS AND SOCIETY

The everyday lives and activities of a society are largely neglected by mathematics education researchers as sites for mathematical knowledge development and use (O'Donoghue, 2000). Many forms of mathematics exist in society side by side with the academic. The forms of mathematics that are embedded in cultural activities, i.e.

everyday mathematics such as street marketing and house building, also fall under the category of ethnomathematics (D'Ambrosio, 1993). Moschkovich & Brenner (2002) defined everyday mathematics as those mathematical practices that adults or children engage in. Bishop (1988) identified six basic activities which can be referred to as "everyday mathematics": counting, locating, measuring, designing, playing and explaining. These out-of-school practices help shed light on ways to improve the teaching and learning of in-school mathematics. Pinxten (1987) noted that if mathematics teaching is founded and built upon the local cultural knowledge and/or cognitive background, it can be more effective and will yield more educational opportunities to marginalized groups. In the following sections we have identified some common everyday math activities from Tamil Nadu and Kerala, two southern states of India. Activities of these sorts can be used in designing curriculum and/or teaching strategies (Moschkovich & Brenner, 2002). The activities mentioned below include mathematical ideas that involve number, logic, spatial configuration and the combination of these into systems or structures. The mathematical concepts embedded in these every day activities will help students understand and experience classroom mathematics in a meaningful way.

NUMBER COUNTING

Eni bi eni" "eji bi eji" "eta n tagba" etc. "eni" represents one, eji represents two, and "eta" represents three and so forth. This is predominant in western cultures even until today. Other countries around the World have special names for number counting.

Number Counting in Yorubaland

Tally, Relics, and Notch

This is originated from writing stroke on the wall to represents number. One stroke represents one, two stroke represents two, three strokes represents three, four strokes represents four, four strokes and fifth stroke drawn across the previous marks four strokes represents five then additional one stroke to that gives six and so on and so forth. It teaches addition of number, as the strokes increases the number progresses. It is used across Africa since late Stone Age till today. E.g. Nigeria, Zimbabwe, Serria lone and another part of the world

Figure 1.

	Numeral	Objects*		Tens**	
1	ení, ọ̀kan	oókàn	11	ọ̀kanlá, oókànlá	10+1
2	èjì	eéjì	12	èjìlá, eéjìlá	10+2
3	ẹ̀ta	ẹẹ́ta	13	ẹ̀talá, ẹẹ́talá	10+3
4	ẹ̀rin	ẹẹ́rin	14	ẹ̀rinlá, ẹẹ́rinlá	10+4
5	àrún	aárùn	15	ẹ̀ẹ́dógún***	20−5
6	ẹ̀fà	ẹẹ́fà	16	ẹẹ́rìndílógún	20−4
7	èje	eéje	17	eétàdílógún	20−3
8	ẹ̀jọ	ẹẹ́jọ	18	eéjìdílógún	20−2
9	ẹ̀sán	ẹẹ́sàn	19	oókàndílógún	20−1
10	ẹ̀wá	ẹẹ́wàá	20	ogún, okòó	

Finger Counting

This is using fingers of the hands and legs in counting as in tally system. It is the act of counting along one's fingers. In African and most cultures, index finger is used to indicate one, index finger and middle finger indicate two and if together with ring finger indicate three and so on. Finger counting serves as a form of manual communication in market place trading, hand signal during open outcry in auction or floor trading.

Ayo Olopo (Game)

"Ayo" game is popularly called "ayo olopon" in the culture of Western Nigeria; other cultures have their special names across African continent. It is a game played by two persons facing each other. It is a wooden board with two rows and six pockets or holes on each row. Each player at a time plays four "ayo" seeds or stones in each of the six pockets or holes. Each of the players takes turn to pick up all the seed or stones in a hole and dropping them one by one round into another holes. Mathematical operations of addition, subtraction and manipulative skills are needed to win the opponent.

Figure 2. Finger counting

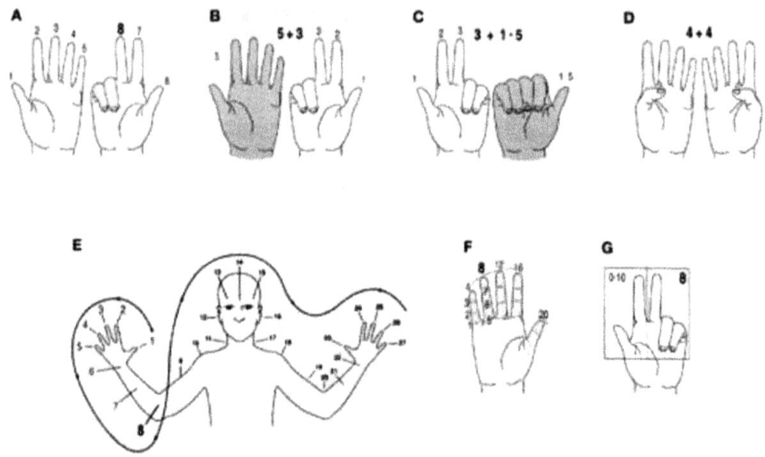

Figure 3. Ayo Olopon

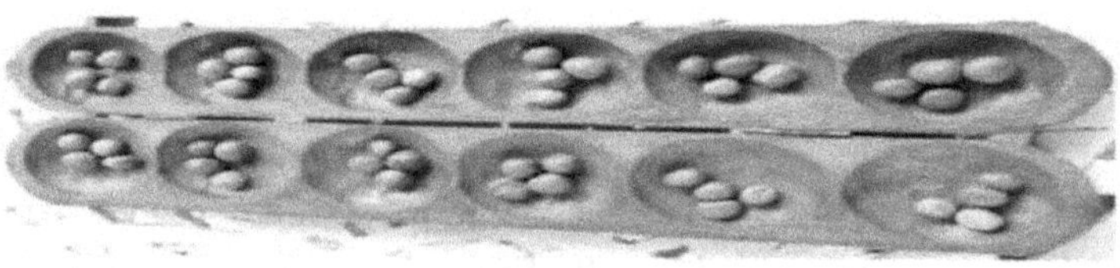

"IJAKADI" (Wrestling)

"Ijakadi" is a game of wrestling. Wrestling is a combat game involving two opponents against each other in grapping type techniques, throwing each other; take downs, joint locks, pins down and so on. The aim is for one to gain and maintain a superior position over the other like a total-no-movement-lock-down, put down position and rising over, superior pin down position and so forth.

1. The opponent who is able to:
2. Technically use his intuition faster to give a pin –fall
3. Pre-calculate implication of opponent combative moves
4. Make a counter-balance move to neutralize and out-play his opponent strength or weak areas in skill wins the game.

Figure 4. Ijakadi

It is a highly intellectual activity involving pre-calculation, intuitive, measurement, pre-empty, summary implication fast thinking and fast acting moves to win the opponent. It is common practice among African culture. The games mentioned above are mostly played by adults. Children of nowadays like online games ranging from game consoles, play station, to mortar combat. In the past children congregated outdoor to play their favorite games freely and did not require much equipment or even none. The games are fun to them, build important skills in them, promote social interaction among children, help the children in body co-ordination, primate strategic thinking, and originate basic Mathematics skills in them. Some of other games that are mostly played by children are:

Ten-Ten

It is a game mostly played by young girls. It is usually played by two persons, facing each other, clapping and raising their legs simultaneously rhythmically. The rule is not to raise the leg directly opposite the leg which her opponent raised. That is if my opponent raises a left leg, then I should raise a right leg and not the other way round. The player who raises the wrong leg gives the other player a point against herself. It rhythmically movement can get so fast with high speed requiring fast thinking, fast acting, high level of self-articulation and agility.

Figure 5. Ten-Ten

Pottery, Weaving, and Folktales

Pottery is the craft of making pot, plates, and cup dish wares out of clay, stones or porcelain. Pottery was developed out of the daily need in day-to-day activities which necessitate fashioning out of different vessel of diverse geometrical shapes for different functional use. Different functional vessels in linear, circular, arcs of two or three-dimensional shapes are formed and improved upon with time.

Weaving is a textile art in making clothes. Two distinct set of yarns or thread are interfaced with each other at right angles to form a fabric or cloth. Early weavers in their cultural setting developed it out of calculated interlocking of materials which can give projective sieve for separating materials, protective covering and adornment for men, wall design fabric or interwoven materials to form a container, cage and so forth. Basket weaving is the use of pliable materials like raffia or bamboo to weave two or three-dimensional domestic vessel or artifact like mat, basket, different containers, bird cage and so forth. They are made with equal spacing and crossed pattern according to the functional use for the particular basket, cage, mat or artifact. The dimensional ordered pattern show its ethno mathematical ideas or thinking behind its fashioning.

Woven designs are used to make cloth, carpet, artistic drawings on walls. It is woven depicting circular, square, cylindrical etc. shapes in a symmetrical arrangement. Local objects and planes are also used in weaving design but design symmetrically to be orderly and good representation of reality. Crowe (2004) shows woven cloth may exhibits one of the seventeen kinds of plane geometry groups. Woven design in most cultures is a local skill in folk art but it has contributed so much to ethno mathematical ideas and practices.

Mathematics Riddles

Classroom teachers in Nigeria can use "everyday math" activities, games or puzzles for family math activities. Through this type of family math activity, children will get an opportunity to communicate mathematics to their parents. Parents will also get to share their experience and knowledge about the mathematics they use every day. If parents get an opportunity to broaden their view of mathematics and mathematics learning, through "everyday mathematics activities", their outlook about modern mathematics may change. This change may also encourage them to get more involved in school related activities. Students can be asked to note down any mathematics they use while they talk about or solve these riddles, puzzles or games. This is an excellent way of asking students to reflect on what they learn in their classrooms. Abreau (2002) noted that research in out-of-school activities suggest that children might benefit from approaches that help them to bridge the gap between school mathematics and what they observe and learn outside school. Additionally, teaching mathematics that builds up from activities like ten-ten and weaving which are considered as "feminine activities," can validate the experience of girls and women and contradict the unfounded notion that they are not able to do mathematics, particularly when involving spatial perception. School mathematics rooted in one's culture, practices and history (Lerman, 1994) could reorient the whole context of mathematics education and allow the learner to experience mathematics as a human rather than an abstract activity.

Textilization, Patterns, Tie, and Dye

African textiles are the major form of expression that Africans use to define themselves. They have used cloth not only for personal adornment but also as a powerful medium of communication for many centuries. Their importance has often been overshadowed by traditional sculpture and masks but in this day and age, we see how African textiles have become the most significant medium by which contemporary African artists are illuminating the connections and continuities between past and recent modes of African artistic expression. African textiles are also a means for us to acquire insight into the social, religious, political and economic complexities of many African communities whose sophisticated cultures we may otherwise remain ignorant about (CAA, 2017).

Textile

Patterns

Pattern making or pattern cutting is the science of designing patterns. A pattern in sewing is the paper or cardboard template from which the parts of a garment are traced onto fabric before cutting out and assembling (sometimes called paper patterns). A custom-fitted basic pattern from which patterns for many different styles can be created is called a sloper (home sewing) or block (industrial production). The flat-pattern method begins with the creation of a sloper or block, a basic pattern for a fitted, Jewel-neck and narrow skirt, made to the wearer's measurements. The sloper is usually made of lightweight cardboard without seam allowances or style details. Once the shape of the sloper has been refined by making a series of mock-up garments called the final sloper can be used in turn to create patterns for many styles of garments with varying necklines, sleeves, dart placements, etc. (Wikipedia, 2017).

Figure 6.

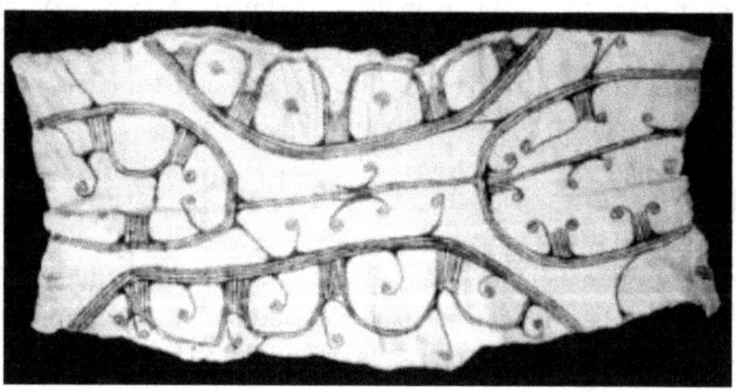

Figure 7.

African Pattern

Tie-dye can be used to create a wide variety of designs on fabric, from standard patterns such as the spiral, peace sign, diamond, and the marble effect to beautiful works of art. Using techniques such as stencils (a la screen printing using dyes or discharge pastes), clamped-on shaped blocks, and tritik (stitching and gathering), tie-dye can produce almost any design desired. Tie-dye techniques have also been used for centuries in the Hausa region of West Africa, with renowned indigo dye pits located in and around Kano, Nigeria. The tie-dyed clothing is then richly embroidered in traditional patterns. It has been suggested that these African techniques were the inspiration for the tie-dyed garments identified with hippie fashion (Hoffmann, Frank, & William, 1994).

CONCLUSION

Bishop (1992), referring to research in mathematics education, argued that similarity is a feature of many research traditions evolving in different countries around the globe. Hence research questions, methods, practices and publications are becoming more standardized. He noted that these similarities should not be taken to mean that there is a universal acceptance of particular research methods or paradigms. We concur with this advice. It is to be expected that in mainly English speaking

Figure 8. A jau woman from the dyers' caste prepares a cotton band for tie-dye colouring, Sangha, Mali, 1980

Anglo western cultures, such as those found in Nigeria and South Africa, that most resources used by our colleagues would be from within this cultural basis. Clearly the focus of much of our research and teaching is within our own culture. However, it is a different thing if we uncritically view, for whatever reason, that what we practice and believe is, or indeed must be, universal or perhaps the best possible option, no matter what the cultural context. In this study we have found a divergence of views that our colleagues hold across a range of issues. However, there does seem to be a trend for colleagues who have experienced at some depth non western cultures to appreciate more deeply some of the issues thrown into relief by the notions of globalization. There also seems to be a wish for global collaboration, as has been noted in the focus group data. We believe that more discussion on the notions of globalization and how it is impacting on mathematics education will challenge us all and give more depth to our thinking and practice. This in turn will position each of us to be better able to deal critically with fundamental issues in the globalized world in which our students and we now live.

Within standardized, broadcast math curricula, concepts and teaching practices are largely dissociated from the knowledge and skills a child brings into the class room. Unless learners realize that mathematics exist in their very own world, beyond school walls and beyond a Eurocentric world view, many of them will continue to complain about it as boring and uninteresting. Universalizing the curriculum for the sake of simplifying assessment or selling textbooks is not going to minimize the anxiety or even hatred that many students feel towards numbers. Conversely, introducing everyday mathematics into curricula will not only help students understand that math is something related to their culture, but will also help them make connections between familiar everyday activities and academic mathematical practices. As well, it will help create a bridge between parents and school, which is a very considerable factor in student success.

Emphasis on the socio-cultural nature of mathematical knowledge also supports a move away from strictly occidental forms of mathematics teaching and opens the door for more active (e.g.; hands on, project-based) and constructivist forms of learning.

REFERENCES

Anderson, S. E. (1990). Worldmath curriculum: Fighting euro-centrism in mathematics. *The Journal of Negro Education*, *59*(3), 348–359. doi:10.2307/2295569

Ascher, M. (1991). *Ethnomathematics: A multicultural view of mathematical ideas*. Pacific Grove, CA: Brooks/Cole Publishing Company.

Atweh, B., Clarkson, P., & Nebres, B. (2003). Mathematics education in international and global context. In *Second international handbook of mathematics education, Part one* (pp. 185–227). Dordrecht: Kluwer Academic Publishers. doi:10.1007/978-94-010-0273-8_7

Bishop, A. J. (1988). *Mathematical enculturation: A cultural perspective on mathematics education*. Dordrecht: Kluwer Academic Publishers. doi:10.1007/978-94-009-2657-8

CAA. (2017). *African textile*. Contemporary-african-art.com.

Clarkson, P., & Atweh, B. (2004). *More Perspectives on the Impact of Globalisation on Mathematics Education in Higher Education in Australasia*. Australian Research Council Large Research Grant.

D'Ambrosio, U. (1993). Ethnomathematics, the nature of mathematics and mathematics education. In P. Ernest (Ed.), *Mathematics, education and philosophy: An international perspective* (pp. 230–242). London: Falmer Press.

de Abreu, G. (2002). Mathematics learning in out-of-school contexts: A cultural psychology perspective. In L. D. English (Ed.), *Handbook of international research in mathematics education* (pp. 323–353). Mahwah, NJ: Lawrence Erlbaum Associates, Inc.

Dowling, P. (1998). *The sociology of mathematics education: Mathematical myths/ pedagogic texts*. London: Falmer Press.

Ernest, P. (2007). *Globalization*. Ideology and Research in Mathematics Education.

Gates, P., & Vistro-Yu, C. P. (2003). Is mathematics for all? In *Second international handbook of mathematics education, Part one* (pp. 31–73). Dordrecht: Kluwer Academic Publishers. doi:10.1007/978-94-010-0273-8_3

Gerdes, P. (1997). On culture, geometrical thinking and mathematics education. In A. B. Powell & M. Frankenstein (Eds.), *Ethnomathematics: Challenging eurocentrism in mathematics education* (pp. 223–247). Albany, NY: State University of New York Press.

Hoffmann, F. W., & William, G. B. (1994). *Fashion & merchandising fads*. New York: Haworth Press.

Joseph, G. G. (2000). *The crest of the peacock. Non-European roots of mathematics*. Princeton, NJ: Princeton University Press.

Lerman, S. (1994). Cultural perspectives on the mathematics classroom. Dordrecht: Kluwer Academic Publishers.

Moschkovich, J.N., & Brenner, M.E. (2002). Preface: Everyday and academic mathematics in the classroom. *Journal of Research in Mathematics Education Monograph*.

O'Donoghue, J. (2000). Adults, mathematics, culture and society. In D. Coben (Ed.), *Perspectives on adults learning mathematics* (pp. 101–107). Dordrecht: Kluwer Academic Publishers.

Pinxten, R. (1987). Applications in the teaching of mathematics and sciences. In A. B. Powell & M. Frankenstein (Eds.), *Ethnomathematics: Challenging eurocentrism in mathematics education* (pp. 373–399). Albany, NY: State University of New York Press.

Rampal, A. (2003). Counting on every day mathematics. In T. S. Saraswati (Ed.), *Crosscultural perspectives in human development: theory, research, and applications* (pp. 326–353). New Delhi: Sage Publications.

Smith, D. (2000). The specific challenges of globalization for teaching and vice versa. *The Alberta Journal of Educational Research*, *46*(1), 7–26.

Varghese, T., & McCusker, D. P. (2006). On Globalization and Ethnomathematics. *Canadian and International Education /Education canadienne et international, 35*(1), 2. Available at: http://ir.lib.uwo.ca/cie-eci/vol35/iss1/2

Section 2

Research on Mathematics Education

Chapter 5
Digital Games for Teaching and Learning of Mathematics in Nigeria Schools

Ayotola Aremu
University of Ibadan, Nigeria

Adebowale Adebagbo
Yaba College of Technology, Nigeria

ABSTRACT

Fractions are among some concepts in mathematics that pupils in Nigeria primary find difficult to learn, and teachers find difficult to teach. These problems require solutions because mathematics knowledge is germane for technological advancement and national development. Pupils need to be actively involved in constructing their knowledge and digital mathematics learning games could be helpful. Thus, this study investigated the effects of digital mathematics games on primary pupils' achievement in fractions concepts. An appropriately validated pupil fractions achievement test (PFAT) was used to collect data for the study. Data collected were analyzed using t-test. It was revealed that pupils exposed to digital mathematics games performed significantly better than those exposed to conventional method. There is significant difference between achievements of male and female pupils exposed to game-based strategy.

DOI: 10.4018/978-1-5225-6158-3.ch005

INTRODUCTION

In order to compete effectively in today's global, information-based economy, and in today's increasingly high-tech work environment, students must be able to reason logically, solve problems and communicate effectively. Though mathematics, science and technology are commonly regarded as "critical filters" for determining future success, satisfactory achievement is not yet obtainable in most students. To prepare all students for twenty-first century careers, increase the number of high achieving students as well as improve on the ability of low achievers in mathematics and to ensure sustainable interest in the subject, students should be provided with innovative opportunities to learn more mathematics.

Students need to be engaged and challenged for the desirable outcomes in learning mathematics hence problem solving ought to be a central focus of all mathematics instruction. To accomplish this, we need to involve them in hands-on activities, to provide them with settings where they can participate in mathematical discovery, to decrease the focus on repetitive tasks, to make available alternate ways of learning concepts, and to offer them activities which they recognize as meaningful. As students benefit from a classroom environment in which they are working to find solutions to meaningful problems, it is also necessary to accord importance to issue of gender.

There are diverse reasons for gender differences in mathematics as this issue can be approached through various perspectives; while many people opine that it is because of preconceived notions that girls will fail and genetic theories, some others say it is due to adult influence, and the list goes on. Even, "math is for boys" stereotype has been used as the definition of why girls do not pursue science and mathematics as their careers (University of Washington, 2011). Also, the "math is not for me" is a cultural stereotype on girls which can affect which career choices and classes they choose to take (University of Washington, 2011). The latter portrays feminine gender as being less confident compared to masculine gender in mathematics abilities, which attributes to why there are so few women pursuing careers in mathematics, science and technology.

However, as quest for solution to the problems of learning mathematics waxes stronger, Haramati (2000) opines that guarantee for real learning is the creation of joy, excitement, and love for learning and these can be created through games as a way of learning. Hence, to improve student interest and academic performance, educators have begun paying special attention to computer games (Oblinger, 2004;

Gee, 2005). The computer games are also called digital games. Some studies (such as 1992; McFarlane, Sparrowhawk & Heald, 2002; Vogel, Vogel, Cannon-Bowers, Bowers, Muse, & Wright, 2006; Ke, 2008; Annetta, Mangrum, Holmes, Collazo, & Cheng, 2009, among others) have been conducted to explore the effects of computer games on student achievement. However, there is no consensus on the effects of digital or computer games in learning mathematics.

MATHEMATICS IN NIGERIA SCHOOLS

Mathematics plays an integral role in art, music, games, explorations, inventions, and commerce within virtually any culture. People in all societies have devised their own ways of doing mathematics consciously and unconsciously ranging from traditional farmers to modern scientists who are involved in one activity or the other, even the women in the kitchen apply reasonable knowledge of mathematics in adding ingredients together and timing of activities in cooking or catering. Often, taking Nigeria as an example, a few pupils are successful in the discipline in schools and there is a phobia for the subject in the society. Lending credence to this report on Nigeria, Salman (2005) confirmed that pupils dislike certain topics in Mathematics because of the belief that they are difficult. Yet, science evolved with the use of mathematical principles and Mathematics is a necessary tool needed by individuals to be able to function effectively in the present technological age.

In Nigeria, a lot of Mathematics educators and researchers like Ohuche, 1978; Akpan, 1987; Oshibodu, 1984 and 1988; Alele-Willaims, 1988; Ale, 1989; Odogwu, 1994; Aremu, 1998; Tella, 1998; Adegboye, 1999; Akinsola, 1999; Amobi, 2006; Adeyemo & Adetona, 2007 and Ugboduma, 2008 cited in Akinoso, 2012), have over the years carried out researches on factors that are responsible for poor performance in mathematics at primary and secondary schools. Some of these include shortage of qualified mathematics teachers, poor facilities, equipment and instructional materials for effective teaching, use of traditional chalk and talk methods, large pupils to teacher ratio, mathematics fright or phobia, student and teacher characteristics (e.g self-efficacy, interest, attitude, gender, family background, socio-economic status, qualification and experience) and lack of activity-based teaching method, to mention but a few.

In spite of these numerous researches and laudable efforts over the years on ways of improving the teaching and learning of Mathematics at primary and secondary school levels, the performance of students at the West African Senior Secondary School Certificate Examinations (WASSCE) is yet to improve significantly as students in Nigeria are still performing poorly in the subject.

Table 1.

Year	Number of candidates	A1-C6		D7-E8		F9	
		N	%	N	%	N	%
1999	756,680	138098	18.3	212514	28.0	381029	50.4
2000	634,604	208244	32.8	196080	30.9	230280	36.3
2001	1,023,102	373955	36.6	334907	32.7	314240	30.7
2002	908,235	309409	34.1	308369	34.0	290457	32.0
2003	926,212	341928	36.9	331348	35.1	229878	24.8
2004	832,689	287484	34.5	245071	29.4	300134	36.0
2005	1,054,853	402982	38.2	267600	25.4	363055	34.4
2006	1,149,277	472674	41.1	357325	31.1	286,826	25.0
2007	1,249,028	584024	46.3	333,844	26.7	302774	24.2
2008	1,268,213	726398	57.3	302266	23.8	218618	17.2
2009	1,348528	634382	47.0	344635	25.6	315738	23.4

Source: Akinoso (2012)

As reported by Azuakola (2012), according to the head of the West African Examination Council, National office, Iyi Uwadiae, candidates with credits in five subjects and above, including English language and Mathematics were 649,156 (38.81) in (2012) out of 1,672,224 candidates; compared with 471,474 (30.61%) in 2011 out of 1,540,250 candidates; and 337,071 (24.94%) in 2010 out of 1,351,557 candidates. Furthermore, quoting Charles Eguridu, the head of the West African Examination Council, National office, Kukogho (2015) reported that only 616,370 candidates (38.68%) of the 1,593,442 total candidates, had credit passes in at least five subjects, including English language and Mathematics. He added that in 2014, only 529,425 candidates (31.28%) had credit passes in at least five subjects, including English language and Mathematics, as against 639,760 (38%), in 2013 examination results. All these reports show poor performance (below 40%) in Mathematics.

To an extent, government at various levels has made concerted efforts to re-orient students' views about studying mathematics, through demystifying mathematics and motivating the interest of all categories of students. From the onset, this is why the Federal Government of Nigeria (FGN, 2004) presents Mathematics as a compulsory subject in the curriculum of the primary and secondary school levels in the Nigeria educational system (FGN, 2004) and a prerequisite to the study of science courses in higher institutions.

COMPUTER GAMES IN TEACHING AND LEARNING

Computer games are also known as digital games and it is any sort of game that is played using a computer. Computer games encompass a broader range of games, including text-based adventure game and other interactive fictions; hence computer games are not necessarily video games. Computer games are resourceful materials in teaching and learning. Learning could be described as a complex process due to the complexity of human nature. Different people have different perceptions and process information differently (Miller, 2001). Different students have different learning styles, which imply that different teaching styles or methods are required to be used by the teachers to meet up with different needs of the students.

Really, to meet the needs of the learners and to improve students' performance having various learning styles, it will be beneficial to explore diverse learning opportunities which can address the intricacies of learners in the teaching-learning activities and to address pupils' varying interests from slow to fast learners. One of the ideal tools for achieving this according to Holland, Jenkins & Squire (2003) is games. They are the ideal tool for accommodating multiple learning styles. McFarlane, Sparrowhawk & Heald, (2002) distinguished three potential uses of computer games in a school environment: general cognitive abilities and skills; affective and motivational aspects; and, knowledge and content-related learning. The strongest empirical support is for the claim that games improve affective and motivational aspects, whereas the claim that the games enhance content-related learning has mixed support (Vogel et al. 2006).

USE OF MATHEMATICS GAMES IN NIGERIA SCHOOLS

The increasing availability of computers has opened up a new world of possibilities forenriching experience that can be employed effectively for teaching and learning particularly with availability of internet. A country like Nigeria with rapidly growing internet revolution is supposed to be at an advantage in the use of computer based tools, such as educational games but this has not been so because, computers have not been practically integrated in the curriculum at any level of our educational system.

A review of literature lends support to the claim that some knowledge domains are particularly suited to computer gaming. Furthermore, empirically, math drill-and-practice games have also been found effective in educational environments (Rice 2007). Also, Chuang & Chen (2009) show that the computer-based video games can promote students' learning and performance. In spite of some studies by Nigerian researchers (Aremu, 1998; Animashahun, 2002; Adebagbo, 2004; Kosoko-Oyedeko, 2012) that report important benefits of Mathematics games, no appreciable effort is being seen made by the government on the use of Mathematics games as basic instructional material or resource in Nigerian secondary and primary schools. This is due to lack of political will by the government and policy makers to associate with, make use of and publicize research findings in Nigeria educational institutions.

Besides, many schools and teachers in Nigeria seem to be ignorant about the utilization of Mathematics games. Azuka (2001) reported that in Nigeria, many Secondary school Mathematics teachers have not been exposed to Mathematics games, not aware of the reasons for playing Mathematics games, do not know the guidelines for preparing Mathematics games, do not know the guidelines for evaluating the usefulness of Mathematics games and there is no period for playing mathematics games in virtually all Secondary Schools. This situation calls for serious attention among the stakeholders. The first must however be to research which games would be effective and consequently can be integrated into the curriculum.

DIGITAL GAMES AND GENDER

Being a key research area, the review of literature on the influence of gender on gaming, points to a number of physical, cognitive and social gender-related differences. Some believe that girls are less likely to enjoy game-play situations that emphasize violence and competition because "they gained a lesser sense of control than they did in other play activities" (Lucas & Sherry, 2004). Also, research on computer use has shown that games are more popular with boys, who are more frequent, intensive and experienced game players than girls (Cassell & Jenkins, 1998; Downes, 1999; Facer, 2003; Harris, 1999 and Mumtaz, 2001).In addition, Griffin (2007a) highlights the distinctiveness of women's interaction with games quoting Lanza in Herz (1997) claiming "Girls are looking for experiences and boys are looking for bragging" (Herz, 1997). Meanwhile, it has been posited that placing educational content in a gaming context would be unfair to women, who play games less and do not enjoy computer gaming as much as men. However, many recent studies have shown no effect for gender (Oyen & Bebko, 1996; Armory, Naicker, Vincent & Adams, 1999; Blumberg & Sokol, 2004 and Ke, 2008) and have shown that while women are less likely to play sports games or violent adventure games they are just as likely as their male counterparts to be playing collaborative massively multiplayer online games (MMOGs).

One study found that women over forty spent more time playing online games than any other group in their study and often stayed up late into the night playing (Sun, Ma, Bao, Chen & Zhang, 2008). Research is showing that women enjoy collaborative games more than men and that it is these games that provide the best avenues for learning in many respects (Ke, 2008). However, findings on gender-specific motivation or performance in computer gaming environment are still inconclusive in all the various categories of computer games.

STATEMENT OF THE PROBLEM

Though mathematics predominantly permeates many aspects of the society, there is inherent phobia for the subject irrespective of status, and this cuts across all generations. The attendant alarming poor performance by the students in mathematics could be associated with a lot of factors such as teacher characteristics, student characteristics, environmental factors, and also the method of instructional delivery, but not limited to these. Moreover, considering the abstract nature of the subject, the struggle with mathematics is much centred on the method of teaching. Hence, the need for novel strategy that could awaken and sustain learner interest is desirable. Nevertheless, game based learning strategy, which is activity based, has the necessary potentials to address the problems of learning and teaching mathematics and which would provide motivation for the students. Besides that most teachers are alien to mathematics games use for teaching. In the teacher training program in Nigeria, such a method of teaching is not given prominence. Thus, there is need to investigate effectiveness of this strategy (game based strategy) for teaching and learning. It is on this premise that this study would examine digital games for teaching and learning of mathematics in Nigeria schools.

Hypotheses

The following null hypotheses were formulated and tested at 0.05 level of significance.

1. There is no significant difference in mathematics achievement of pupils that are exposed to computer games strategy and those exposed to modified conventional method in solving problems on fractions concepts.
2. There is no significant difference in mathematics achievements of male and female pupils that are exposed to computer games strategy in solving problems on fractions concepts.

Research Design

This study adopted a quasi-experimental research design with the use of pretest-posttest control group quasi experimental research design.

Sampling and Sample

The population for this study consisted of all pupils in public and private primary schools, in all the Local Government Education Areas (within the six Education Districts of the 20 Local Government Areas) of Lagos state, Nigeria. In selecting sample for this study, a stratified random sampling technique was used according to the six educational districts; then, purposive method was used to select three districts with functional computer laboratory facilities from which purposive method was employed in selecting two representative schools that were randomly assigned to experimental and control groups.

Instrumentation

Only one instrument was used in the study; the Pupil Fractions Achievement Test (PFAT). The PFAT is a fifteen (15) multiple-choice test with options ranging from A to D. It was designed to measure pupils' achievement in solving fractions problems. Then, necessary face and content validities were conducted. After these, Kudar Richardson formula 21 (K-R 21) was used for reliability computation and its reliability coefficient was 0.86.

Game Description

The digital game used in this study was an offline version of free open source software, produced by Jamit software, Jamit Fraction Games. The game was adapted, made to YoungMind Games, to reflect Nigeria environment by replacing some graphics and objects that were considered foreign. The concept of the game was that an avatar, which does not portray any gender characteristics, wants to climb a staircase to get a bunch of banana. Every fraction problem solved by the game player will make the avatar to jump one step up on the staircase while wrong answer will bring it down the staircase. A monkey is also competing to get the banana before the avatar can get there. The faster the player is able to get the avatar to the top of the staircase, the better, else the monkey will take the banana. The game will be won by the player if avatar gets the banana but game lost if the monkey gets the banana. This game is played individually; also, the game has different in-built supports and scaffoldings to solve any learning challenges.

Figure 1. Jamit fraction games shot

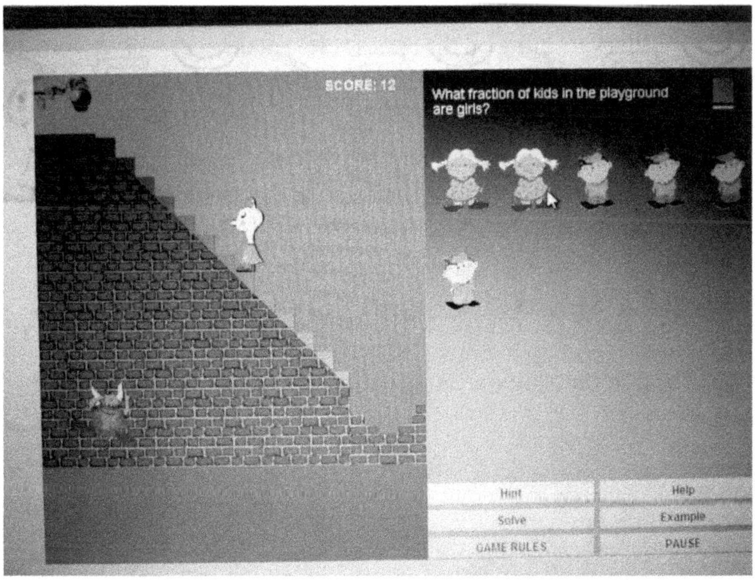

Figure 2. YoungMind Games shot

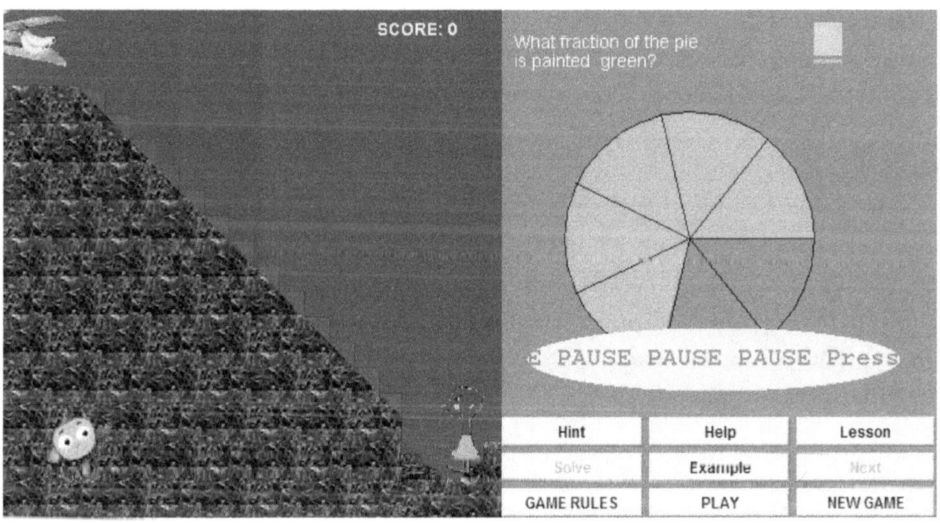

Procedure for Data Collection

Necessary preliminary training on the use of relevant devices was given to teachers in the experimental group appropriately. This was then followed by a pre-test conducted same day in all the participating schools. The experimental group was exposed to game based learning strategy treatment while conventional method was used for the control group. Immediately after the experiment treatment, post-test was done the same manner for PFAT.

Data Analysis and Results

The data analyses on the study are presented as follow according to the stated two hypotheses.

Hypothesis 1

There is no significant difference in mathematics achievement of pupils that are exposed to computer games strategy and those exposed to modified conventional method in solving problems on fractions concepts.

From the Table 2, N=25 for each of the groups while the mean of the experimental group is almost the same with the mean of the control group. The t-test analysis also shows that t_{cal} is 0.04 at 48 df which is less than 0.05 alpha level of significance at 0.05 level of significance. Thus, there is no significant difference between the two groups indicating equal levels of entry behaviour for both the experimental and the control groups.

The Table 3 shows that the mean of the experimental group is greater than the mean of the control group while the t-test shows that t_{cal} is 1.19 at 48 df which is greater than 0.05 alpha level of significance. Therefore, the achievement levels of pupils in the experimental and the control groups are significantly different. Hence, the hypothesis 1 is rejected.

Table 2. t-test showing pretest mathematics achievement scores between experimental and control groups

Groups	N	Mean	S.D	df	t_{cal}	Sig.
Experimental	25	5.0	1.7	48	.04	< 0.05 ns
Conventional	25	4.9	1.8			

ns denotes Not Significant at 0.05

Table 3. t-test showing posttest mathematics achievement scores between experimental and control groups

Groups	N	Mean	S.D	df	t_{cal}	Sig.
Experimental	25	10.3	1.7	48	1.19	> 0.05*
Conventional	25	8.8	1.9			

* denotes Significant at 0.05

Hypothesis 2

There is no significant difference in mathematics achievements of male and female pupils that are exposed to computer games strategy in solving problems on fractions concepts.

From the Table 4, the mean of the male pupils is same as the mean of the female pupils in the experimental group; which is 5.0 (for male N=10 and for female N=15) but the t-test result shows t_{cal} = .000 at 23 df which is less 0.05 alpha level of significance. Hence, there is no significant difference between the males and the females in the experimental group. This indicates that there is also no significant difference in entry behaviour of both groups.

The Table 5 indicates that the mean of the female (N = 15) pupils is greater than the mean of the male (N = 10) pupils. However, the t-test shows that t_{cal} = .666 at 23 df which is greater than 0.05 level of significance. This shows that there is significant difference in achievement of the males and the females in the experimental group; hence, the hypothesis 2 is rejected.

Table 4. t test showing pretest achievement scores between male and female experimental group

Gender	N	Mean	S.D	df	t_{cal}	Sig.
Male	10	5.0	1.3	23	.000	< 0.05 ns
Female	15	5.0	1.9			

ns denotes Not Significant at 0.05

Table 5. t-test showing posttest achievement scores between male and female experimental group

Gender	N	Mean	S.D	df	t_{cal}	Sig.
Male	10	9.7	1.4	23	.666	> 0.05*
Female	15	10.7	1.8			

* denotes Significant at 0.05

DISCUSSION OF THE RESULT

In this study, the experimental group performed better than the control group as the former shows a better achievement compared to the control group. This is contrary to the findings of Miller, Lehman & Koedinger (1999) that students may be distracted by game-playing, and thus, not achieving the learning goals. This present finding also contradicts the report of Ya-Ting (2012) that found no impact of games on academic achievement. However, it supports the findings of Gee, (2003); Chuang & Chen (2009); Yang, (2012) and Hwang, Wu and Chen, (2015) that the computer based video games can promote students' learning and performance; and could support teaching. This implies that game based learning can help pupils to achieve better result in mathematics; particularly if it contains local content that are designed to meet the level of standard educational games. Hence, Nigeria educators and education support institutions should assume the responsibility of designing educational games with local content. In addition to fun and play that could sustain learning interest, this will help the pupils to attain better and improved achievement in learning mathematics.

Moreover, the male and the female pupils in the experimental group improved tremendously in their mathematics achievement. While achievement was almost double by the male, the female overshoot the double level of their achievement. This supports the finding of Oyen & Bebko, 1996; Armory et al., 1999; Blumberg & Sokol, 2004; Vogel et al., 2006 and Ke, 2008 that educational content in a gaming context would not be unfair to women; and they asserted generic benefits for both gender. This current finding corroborates the result of Cooper, 2006 that the use of technology in education affects girls and boys differently. It then follows, based on the findings of this current study, that if the totality of games content, design and technology are neutral to gender, the performance of female pupils would be improved and could be better than their male counterparts thereby bridging the gender gap between the male and the female pupils' achievement in learning mathematics. Hence, game-based learning strategy could improve performance and population of female pupils, who will be interested in learning mathematics; and eventually making career out of the subject to contribute to technological advancement of Nigeria, even the world at large.

CONCLUSION AND RECOMMENDATIONS

The quest of the society to provide a lasting solution to poor mathematics achievement of pupils in Nigeria schools could be met if computer games are resourcefully integrated into curriculum implementation. The reality of computer based educational games is beyond being treated merely as ordinary play devices, rather as productive resources with contemporary way of arousing learner interests, sustaining learning and activities engagement with fun; in a fearless environment where failure is not a threat but achievement or success is voluntarily a burning aspiration for the learners.

The game-based learning strategy, as shown in the study, could help in reversing the situation of poor performance of pupils in mathematics and could close gaps of academic achievement that may be caused by gender. For the next desirable level of development, computer mathematics games should be promoted in teacher training programs and institutions, in order to make game-based learning strategy a major part of teacher training program.

Besides, practicing teachers should be trained on how to produce standard educational games that have local content. Furthermore, government and all relevant stakeholders (such as corporate organisations, professional association, parent's association, old students association, non-governmental organisations, philanthropists, etc.) should provide fund and required support to encourage professional teachers in producing and possibly specializing on standard local content based educational games.

REFERENCES

Adebagbo, A. O. (2004). *Effect of Games on Primary Pupils' Learning Outcomes in Developmentof Multiplication Skills* (Unpublished M.Ed Dissertation). Department of Teacher Education, University of Ibadan, Ibadan, Nigeria.

Adegboye, A. O. (1999). The assessment of laboratory approach in the teaching of mathematics. *Nigeria Education Digest, 1&2*, 149-162.

Adeyemo, D. A., & Adetona, M.O. (2007). A path analytic study of the personality factors affecting students' learning outcomes in Mathematics. *European Journal of Scientific Research, 18*(1).

Akinoso, S. O. (2012). *Effects of Concrete-Representational-Abstract and Explicit Instructional Strategies On Senior Secondary School Students' Achievement In And Attitude To Mathematics*. A Ph.D Post-Field Report Presented At The Joint Staff/ Higher Degree Students Seminar, Department Of Teacher Education, Faculty Of Education, University Of Ibadan, Ibadan, Nigeria.

Akinsola, M. K. (1999). Effect of instruction on students' performance on knowledge comprehension and application tasks in mathematics. *African Journal of Education Research, 5*(1), 94–100.

Akinsola, O. S. (2001). The extent of dominant factors of girls and women's low participation in Mathematics: Mathematics educators' views. *Proceedings of the 42nd Annual Conference of the Science Teachers Association of Nigerian Mathematics Panel, 295-298.*

Akpan, A. A. (1987). *Correlates of Mathematical Problem-Solving Ability among Secondary School Students* (Unpublished Ph.D. Thesis). University of Ibadan, Ibadan, Nigeria.

Ale, S. O. (1989). School Mathematics in the 1990's some Major Problems for Developing Countries. *International Journal of Mathematical Education in Science and Technology, 20*(5), 655–659. doi:10.1080/0020739890200502

Alele, W. G. (1988). Keynote Address Delivered at the Silver Jubilee meeting of mathematics Association of Nigeria (MAN) Sept. 1982. *Abacus, 18*(1), 15–19.

Amobi, F. A. (2006). Beyond the call: Preserving reflection in the preparation of 'high qualified' teacher. *Teacher Education Quarterly*, *33*(2), 23–33.

Animashahun, I. A. (2002). The Effects of Simulation-Games on Students' Achievement in and Attitude to Mathematics (Unpublished M.Ed Project). University of Ibadan.

Annetta, L., Mangrum, J., Holmes, S., Collazo, K., & Cheng, M.-T. (2009). Bridging reality to virtual reality: Investigating gender effect and student engagement on learning through video game play in an elementary school classroom. *International Journal of Science Education*, *31*(8), 1091–1113. doi:10.1080/09500690801968656

Aremu, S. A. (1998). *Effects of card and Geoboard game-based Instructional Strategies on Primary School Pupils' Achievement in Practical Geometry* (Unpublished Ph.D Thesis). Department of Teacher Education, University of Ibadan.

Armory, A., Naicker, K., Vincent, J., & Adams, C. (1999). The use of computer games as an educational tool: Identification of appropriate game types and game elements. *British Journal of Educational Technology*, *30*(4), 311–321. doi:10.1111/1467-8535.00121

Azuakola, S. (2012). *Naija.com*. Retrieved from ynaija.com/better-but-far-from-good-waec-releases-mayjune-2012-ssce-results/

Azuka, B. F. (2001). *The use of Mathematical Games in Nigerian Secondary Schools*. Paper presented at Mathematical Association Nigeria Conference, Katsina.

Blumberg, F., & Sokol, L. M. (2004). Boys' and Girls' Use of Cognitive Strategy When learning to Play Video Games. *The Journal of General Psychology*, *131*(2), 151–158. doi:10.3200/GENP.131.2.151-158 PMID:15088867

Cassell, J., & Jenkins, H. (Eds.). (1998). *From Barbie to mortal kombat: Gender and computer games*. Cambridge, MA: MIT Press.

Chuang, T., & Chen, W. (2009). Effect of computer-based video games on children: An experimental study. *Journal of Educational Technology & Society*, *12*(2), 1–10.

Cooper, J. (2006). The digital divide: The special case of gender. *Journal of Computer Assisted Learning*, *22*(5), 320–334. doi:10.1111/j.1365-2729.2006.00185.x

Downes, T. (1999). Playing with computing technologies in the home. *Education and Information Technologies*, *4*(1), 65–79. doi:10.1023/A:1009607432286

Facer, K. (2003). *Computer games and learning*. Retrieved from http://www.futurelab.org.uk/resources/documents/discussion_papers/Computer_Games_and_Learning_discpaper.pdf

Federal Government of Nigeria. (2004). *National Policy on Education*. Lagos: Federal Ministry of Education.

Gee, J. P. (2003). *What video games have to teach us about learning and literacy*. New York: Palgrave Macmillan.

Gee, J. P. (2005). What Would a State of the Art Instructional Game Look Like? *Innovate (North Miami Beach, Fla.)*, *1*(6).

Griffin, S. (2007a). Early intervention for children at risk of developing mathematical learning difficulties. In D. B. Berch & M. M. M. Mazzocco (Eds.), *Why is math so hard for some children? The nature and origins of mathematical learning difficulties and disabilities* (pp. 373–395). Baltimore, MD: Paul H. Brooks Publishing Co.

Haramati, A. (2000). Teaching physiology: Filling a bucket or lighting a fire? *The Physiologist*, *43*, 117–121. PMID:10916409

Harris, S. (1999). Secondary school students' use of computers at home. *British Journal of Educational Technology*, *30*(4), 331–339. doi:10.1111/1467-8535.00123

Herz, J. C. (1997). *Joystick Nation. How videogames ate our quarters, won our hearts, and rewired our minds*. Princeton, NJ: Little Brown & Company.

Holland, W., Jenkins, H., & Squire, K. (2003). Theory by Design. In M. J. P. Wolf & B. Perron (Eds.), *The video game theory reader* (p. xxii). New York: Routledge.

Hwang, G. J., Wu, P. H., & Chen, C. C. (2015). An online game approach for improving students' learning performance in web-based problem-solving activities. *Computers & Education*, *59*(1), 1246–1256.

Jamit Software. (1999 – 2003). Jamit Fraction Games.

Ke, F. (2008). Computer games application within alternative classroom goal structures: Cognitive, metacognitive, and affective evaluation. *Educational Technology Research and Development*, 56.

Kosoko-Oyedeko, G. A. (2012). Effects of Instructional Games on Primary School Pupils' Learning Outcomes in Christian Religious Studies in Lagos State, Nigeria. A P.hD post-field seminar paper presented to the Department of Teacher Education, University of Ibadan, Ibadan, Nigeria.

Kukogoh, I. S. (2015). *Pulse newsletter 15:09 11:08.2015. Pulse.ng*. WAEC.

Lucas, K., & Sherry, J. L. (2004). Sex differences in video game play: A communication-based explanation. *Communication Research, 31*(5), 499–523. doi:10.1177/0093650204267930

McFarlane, A., Sparrowhawk, A., & Heald, Y. (2002). *Report on the educational use of games*. Cambridge: Teachers Evaluating Educational Multimedia.

Miller, C. S., Lehman, J. F., & Koedinger, K. R. (1999). Goals and learning in microworlds. *Cognitive Science, 23*(3), 305–336. doi:10.120715516709cog2303_2

Miller, P. (2001). Learning styles: The multimedia of the mind. *ERIC Document ED, 451*, 140.

Mumtaz, S. (2001). Children's enjoyment and perception of computer use in the home and the school. *Computers & Education, 36*(4), 347–362. doi:10.1016/S0360-1315(01)00023-9

Oblinger, D. (2004). The next generation of educational engagement. *Journal of Interactive Media in Education, 2004*(8), 1-18.

Odogwu, H. N. (1994). Primary school Teachers and the Teaching of Time Concept in Schools. *Education Today, 7*(21), 24–30.

Ohuche, R. O. (1978). Recent attempts at Mathematics Curriculum Renewal in English Speaking West Africa. *Abacus, 12*(1), 17–26.

Oshibodu, B. M. (1984). An Investigation of Teacher's Instrumental problems in Mathematics. *Benin Journal of Educational Studies., 11*(3), 43–54.

Oyen, A., & Bebko, J. M. (1996). The Effects of Computer Games and Lesson Contexts on Children's Mnemonic Strategies. *Journal of Experimental Child Psychology, 62*(2), 173–189. doi:10.1006/jecp.1996.0027 PMID:8812043

Rice, J. (2007). New media resistance: Barriers to implementation of computer video games in the classroom. 668. *Journal of Educational Multimedia and Hypermedia, 16*(3), 249–261.

Salman, M. F. (2005). Teachers' identification of the difficulty levels of topics in the primary school mathematics curriculum in Kwara state abacus. *The Journal of Mathematical Association of Nigeria, 30*(1), 20–29.

Sun, D., Ma, N., Bao, M., Chen, X.-C., & Zhang, D.-R. (2008). Computer Games: A Double Edged Sword? *Cyberpsychology & Behavior, 11*(5), 545–548. doi:10.1089/cpb.2007.0145 PMID:18817481

Tella, A. (1998). *An Investigation into Poor Performance and Attitude of Secondary School Students towards Mathematics in Oyo State, Nigeria*. Ibadan: Project University of Ibadan.

Ugboduma S. O. (2008). Introducing Mathematics laboratory into secondary schools Mathematics curriculum to promote students' enthusian for natural development. *Journal of Mathematical Association of Nigeria,* 187-196.

University of Washington. (2011). Gender stereotypes about math develop as early as second grade. *ScienceDaily*.

Vogel, J. J., Vogel, D. S., Cannon-Bowers, J., Bowers, C. A., Muse, K., & Wright, M. (2006). Computer gaming and interactive simulations for learning: A meta-analysis. *Journal of Educational Computing Research*, *34*(3), 229–243. doi:10.2190/FLHV-K4WA-WPVQ-H0YM

Yang, T. C. (2012). Building virtual cities, inspiring intelligent citizens: Digital games for developing students' problem solving and learning motivation. *Computers & Education*, *59*(2), 365–377. doi:10.1016/j.compedu.2012.01.012

Chapter 6
Effects of Data-Driven Instructional Strategy on Pre-Service Teachers' Mathematics Lesson Preparation in College of Education Ikere-Ekiti

Oluwabunmi V. Kehinde-Dada
University of Ibadan, Nigeria

Olusola Obisanya
Heigher Heights School, Nigeria

Gbenga Adewale
University of Ibadan, Nigeria

ABSTRACT

This study determined the effects of data-driven instructional strategy (DDIS) on pre-service teachers' lesson preparation in mathematics. The moderating effect of gender was also examined. A pretest-posttest control group quasi-experimental design was adopted. Twenty-seven College of Education pre-service teachers participated in the study. The study was guided by three research questions. Data collected were analyzed using analysis of variance (ANOVA). Instruments used were rating scale tagged practical teaching assessment scale (PTAS) (Scotts' pie = 0.87) and data-driven instructional package (DDIP). There was a significant main effect of treatment on pre-service teachers' lesson preparation ($F(1,23) = 26.83$, partial $\eta 2 = 0.54$). Lesson preparations of pre-service teachers exposed to DDIS ($\bar{x} = 7.12$) were better than those of control ($\bar{x} = 4.87$). There was a significant interaction effect of treatment and teachers' gender on pre-service teachers' lesson preparation ($F(1,23) = 4.37$). Data-driven instructional strategy enhanced pre-service teachers' lesson preparation in mathematics.

DOI: 10.4018/978-1-5225-6158-3.ch006

INTRODUCTION

Improving instruction in mathematics education has been a major topic for teachers, researchers, administrators and the public (Popoola, 2004). Teachers play a crucial role in the educational process because of the enormous responsibilities placed on them for the translation and implementation of curricular contents, instructional materials as well as assessments of students' learning outcomes (Bolaji, 2005). The importance of the teacher in the meaningful development of education at all levels is reflected in the National Policy on Education (NPE) as it declares that no educational system can rise above the quality of its teachers (FRN, 2013). This declaration underscores the need for effective teaching and learning of mathematics from primary schools to tertiary institutions.

Studies carried out by House and Telese (2008) show that the teachers' method of teaching mathematics as well as their personality contributes to students' attitude towards mathematics. Yara (2009) notes that the attitude of students can be influenced by their teachers' attitude and teaching method. Without interest and personal effort in learning mathematics, students can hardly perform well in the subject. Therefore, attempts to improve the teaching and learning of mathematics will be successful to the degree to which the learners' interests and teachers' teaching methods are considered.

Researchers stress the use of multiple data sources, since there is no single assessment that can inform educators all they need to know to make well-informed instructional decisions (Adewale and Ibidiran 2013). Schools collect enormous amount of data on students' attendance, behaviour, and performance, as well as administrative data and perceptual data from surveys and focus groups, but when it comes to improving instruction and learning, it is not the quantity of the data that counts, but how the information is used (Hamilton 2009). In the teaching- learning process, a teacher is expected to organize curriculum content and teaching with the aim of helping children to overcome barriers to learning. There are various methods of improving the teachings of mathematics in secondary schools which are exposed to pre-service teachers in the colleges of education. Some of them are demonstration, guided inquiry, laboratory, individualized instruction, computer assisted instruction, problem solving and team teaching but these have been neglected due to the school system factors. It is imperative at this juncture to introduce a new instructional strategy that can ameliorate or reduce the problem of poor performance of students in Mathematics like data driven instruction strategy.

Data analysis can provide a snapshot of what students know, what they should know, and what can be done to meet their academic needs. With appropriate analysis and interpretation of data, educators can make informed decisions that positively affect students' learning outcomes. This means that without appropriate analyses of students' data in terms of achievement, there is the likelihood that learning outcome might be negative. This is why, it is necessary to use the information gathered about students' learning outcome (this method is known as data diven instruction) to inform decision on the students and use appropriate teaching method to teach them in order to improve their learning outcome. Data-driven instruction is an instructional package that focuses on facts about what students actually learned using data-based methods and not the traditional methods that emphasize what teachers apparently taught in schools.

There are four major aspects of data-driven instruction: culture, assessment, analysis and action (Bambrick-Santoyo 2010). A school's culture includes student aspirations and a code of conduct that promotes positive learning behaviours, which instructions alone might not be able to do without aspirations and a code of conduct. Assessment, analysis, and action are indispensable in increasing students' achievement, and should be included in the school's culture (Christie 2005 in Bambrick-Santoyo 2010). Many school heads use a simple but highly effective yearly data calendar which they display publicly and refer to constantly so that everyone in the school community, including students and their guardians know when important events will take place. In the school culture, the principal has a right to carve out time for the assessment, analysis, and action, and also any professional development teachers need to succeed in each part of the cycle. In order to get everyone in the school working toward the same goal of preparing every child for both internal and external examination and career readiness, teachers must adhere to the same standards and assessments for all students in a given class level and content area (Velom and Rohani 2010). This is very important if teachers are to collaborate on data analysis, professional development, and strategies for re-teaching and improving students' learning.

Many schools sometimes feel a great urgency to improve students' performance. These schools subsequently design assessments without having sufficient time to think through their content. This kind of assessment is a formative one which in this study refers to as interim assessment. For an interim assessment to produce useful information about how students perform in school examinations, each question needs to be clearly aligned to a standard (Bambrick-Santoyo 2010). The questions should reflect both the rigour and format of the standard examinations if they are to guide teachers on how much students are expected to know and be able to do. In this way, assessments clearly define standard and render them meaningful. When selecting or designing an interim assessment aligned to state standards, the researcher works with teachers to drill down to required level of specificity, and goes beyond the state standards. For example, a school that must meet the standard 'plane shapes and calculations', the students' mastery of this standard will depend greatly on whether the student confronts assessments with different types of angles or with complex fractions embedded in a multistep word problem.

The analysis involves focusing on the individual questions in an assessment. According to Yeh (2006) data reports usually provide an overall snapshot of how the students' perform on specific standards. It shows whether students answered an individual question correctly but lacks information on students' wrong responses. This is often referred to as "item analysis" reports. With the data analyzed, researchers and teachers can dig much deeper in the sense that, they can identify the specific challenges individual students or groups of students are facing. This will guide them to pinpoint their errors. Getting to this deeper level of analysis will lead to a focus on subsequent actions in the classroom.

The process of collecting and analyzing student data from assessments is important but only becomes meaningful when the process is combined with effective action from the teacher (Guskey 2005). For effective results, this stage focuses on teachers re-teaching the whole group. It allows teachers to concentrate on a few key concepts that the majority of students are struggling with. It also involves class-level and content-area teachers coming to an agreement on what concepts need to be re-taught, when each topic will be re-taught, and how to determine students' mastery after re-teaching. Agreement on the "what," "when," and "reassessment" plan ensures that standards get implemented consistently with the entire teaching team. The steps will provide teachers with a digestible approach to "action" and allow researcher and school leaders the ability to monitor implementation.

Teacher's lesson plan, often dictates the mode of presentation the lesson will take (Kellough and Kellough, 2007). Daily lesson note preparation can take many forms; the features of a particular lesson often determine the lesson note format (Arends 2004). Lesson preparation and delivery can also determine whether the learning would be effective or not because these dictate what happen during the lesson (Bahr and Garcia, 2010). Some scholars have observed that pre-service teachers in Nigeria are often taught using lecture method, and the lesson format they are exposed to is teacher-centered (Olosunde, 2009; Salami, 2009). Lesson note preparation and delivery skills are necessary skills a teacher must process in order to teach effectively and successfully. Hence, there is a need to focus on how to develop the lesson note preparation and delivery skills on data-driven instructions in pre-service teacher training institutions so as to make the teacher more effective, especially those teaching important subject like mathematics that students do not always like. The Practical Teaching Assessment Rating Scale used currently in University of Nigeria, Nsukka- UNN (the samples used in this study were drawn from one of the UNN affiliates) has the following features:

1. General information on the class and subject.
2. Specific statement of objectives (as reflecting facts/ knowledge, values/ attitudes and skills)
3. Content (logical, properly sequenced)
4. Adequacy

Any lesson delivered using this format is mostly student-centred. In some developed countries like America, Europe and China, where effective teaching is emphasized in the school system, and different strategies of teaching with lesson plans format for lesson preparation data-driven instruction that could enhance the implementation of such strategies is used (Bambrick-Santoyo 2010). Data-driven instruction of strategy has a six part model which includes:

1. Pre-assessment, where the entry behaviour of the students will be measured
2. Selection of topics to be taught based on the difficult areas from the simulated group
3. Equipment of relevant skills for teaching using data (teaching methods)
4. Giving of interim assessment to determine the learning outcomes and
5. Re-teach until at least 80% of the student improve
6. Evaluate and analyse data as feedback on the effectiveness of the instruction.

Special teaching method is a course that all second year, mathematics' students of University of Nigeria Nsukka in affiliation with College of Education Ekiti State offer and which launch them out for teaching practice. It is a method of teaching which explains what strategy and what one must do before adopting any strategy of teaching. It tells about factors that may influence ones' choice of method of teaching and also different types of teaching methods.

Statement of the Problem

Research findings revealed that a large number of junior secondary school mathematics teachers trained in the colleges of education in Nigeria find it difficult to develop lesson notes and also prepare for lesson. Past studies on data-driven instructions have focused on the general effectiveness of the strategy on English language with less focus on the pre-service teachers' lesson preparations as well as junior secondary school students' learning outcomes in mathematics. Also, most of the available studies looked at methods of teaching that do not involve collecting student-related data that can improve students' learning outcomes in mathematics. This study filled the gap by adopting data-driven instructional strategy in the teaching and learning process with the view of improving pre-service lesson preparation. Effect of teacher gender was also examined.

Research Questions

1. What is the impact of the treatment (special teaching method and data-driven instruction) on Pre-service teachers' lesson preparation?
2. Is there a significant main effect of teachers' gender on Pre-service teachers' lesson preparation?
3. Is there a significant interaction effect of the treatment and teachers' gender on pre-service teachers' lesson preparation?

Methodology

This study adopted a pretest-posttest, control group quasi-experimental design using a 2x2 factorial matrix as shown in Tables 1. The pre-service teachers were divided into two groups. The first group was trained on 'special teaching methods' plus Data Driven Instruction while the second group was exposed to the 'special teaching methods' only.

Table 1. Layout of the 2X2 Factorial Matrix of treatment and gender for the first phase

Pre-service Teachers' gender	Treatment	
	X1 (STM + DDI)	Control (STM)
Male		
Female		

Experiment Structural Design

$Q_1\ X_1\ Q_2$

$Q_1\ X_2\ Q_2$

where

Q_1: Pre - test for each group
Q_2: Post-test for each group
X_1= Special Teaching Method (STM) + Data-Driven Instruction (Experimental group)
X_2 = Special teaching method (STM) + convectional teaching (Control group)

Variables in the Study

1. Independent Variable
   a. STM+DDI (Treatment)
   b. STM (Control)
2. Moderator Variable
   a. Teacher gender (Male & Female)
3. Dependent Variable
   a. Achievement in lesson preparation

The population of the study was twenty-seven students studying mathematics in part two of their programme in College of Education Ikere-Ekiti in affiliation with university of Nigeria Nsukka (full-time). These students are qualified mathematics teachers in junior secondary school undergoing degree programme all the part two students studying Mathematics in the College of Education Ikere-Ekiti in affiliation with University of Nigeria Nsukka (full-time) were involved, a multi-stage sampling technique was used to select participants of this study. Purposive sampling was used to select College of Education Ikere Ekiti based on the condition that the researcher teaches in the college. Purposive sampling was used to select all part two students who offered ED 222 (special teaching method) course because, it is a compulsory course they must take before going for teaching practice; the pre-service teachers were twenty-seven in all. All the pre-service students were taken to a secondary school for observation on lesson preparation (pre-test). Simple random sampling was used to select thirteen pre-service students that were trained data-driven instruction and the remaining 14 on conventional method. After the training, the pre-service teachers were also observed (post-test).

Instrumentation

Two types of research instruments were used for this study; these are response and stimulus instruments.

1. Response instruments
    a. Practical Teaching Assessment Scale (PTAS)
2. Stimulus
    a. Special teaching method
    b. Data Driven instructional Package
    c. Conventional method of teaching

Practical Teaching Assessment Scale (PTAS)

This is a scale adapted from university of Nigeria, Nsukka. It contained sixteen (16) items in the domains of lesson preparation and presentation (delivery). The scale was used to assess pre-service teachers based on the stipulated marks, depending on each items on the scale. The scale was subjected to experts' criticism and comments. The criticism and comments of the experts were used to establish the content and construct validity of the instrument. The reliability was established using inter-rater (scotts' pie) at 0.87.

Data-Driven Instructional Package (DDIP)

This was adapted by the researcher from the work of Bambrick-Santoyo (2010) and it consists of the guide on data-driven instructional strategy as well as the package to be delivered. This was used to prepare the pre-service teachers in the experimental group. The instrument covered the six topics in JSS2 second term scheme of work in Ekiti state.

Data-Driven Instruction Model

The model used in this study is 6-part model suggested by Bambrick-Santoyo (2010). They are:

1. Pre-assessment, where the entry behaviour of the students will be measured
2. Selection of topics to be taught based on the difficult areas from the simulated group

3. Equipment of relevant skills for teaching using data (teaching methods)
4. Giving of interim assessment to determine the learning outcomes
5. Re-teach until at least 80% of the student improve
6. Evaluate and analyse data as feedback on the effectiveness of the instruction. This is represented in Figure 1.

Conventional Method

The same topics taught in the DDIS were exposed to the control group without the use of DDIS. They were taught in the same way other courses were taught in the College.

RESULTS AND DISCUSSION

The results are presented and discussed with respect to the stated research questions. Analysis of Covariance (ANCOVA) was used to examine the research questions stated at 0.05 level of significance ($p < 0.05$).

Figure 1. The steps in Data-driven instruction model, application of Data-driven Instruction in this study and data-driven instruction phase 1 on lesson preparation

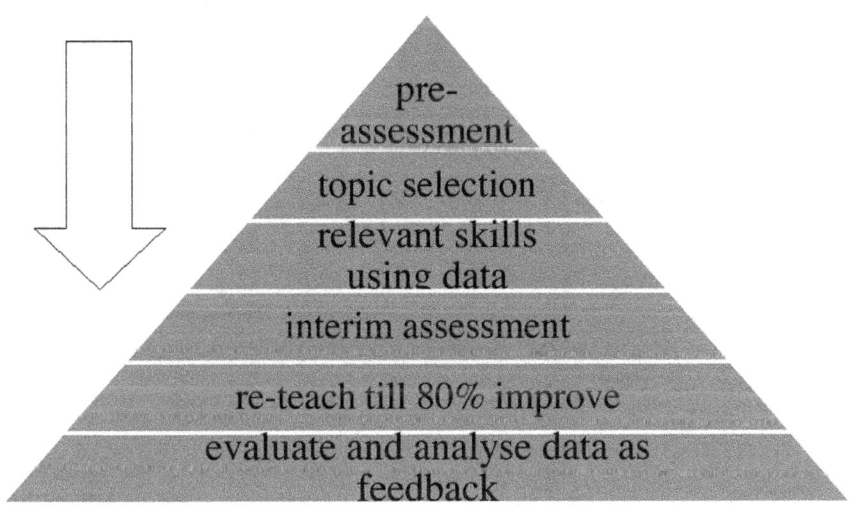

Box 1.

Stage One: Pre-Assessment (Simulation of data)

- Researcher analyzed the result of the past pre-service teachers that have undertaken the special teaching method course to determine:
    i. areas of difficulties in the topic covered and the use of the course during teaching practice
    ii. common mistakes pre-service teachers made during their teaching practices
    iii. use of appropriate model and arithmetic operations during teaching
    iv. use of appropriate and relevant instructional materials
    v. writing style and communication skills

Stage Two: Topic Selection

Using the result of the analysis: Teacher

- Selected different topics based on the severity of difficulty. These formed the topics treated.

Stage Three: Relevant Skills Using Data

The data already generated in stage one informed the teaching skills. For example, it was discovered that students find 'lesson note preparation and delivery' difficult, this was broken down as follows:

i. Subject: Mathematics
ii. Class: JSS2
iii. Topic: Algebraic expansion
iv. Content: Removing brackets, expanding algebraic expressions
v. Ref bk:NGM book2, Essential mathematics
vi. Objective: By the end of the lesson, students should be able to simplify and expand algebraic expression
vii. Previous knowledge: students are familiar with Mathematics signs such as +, -, x, ÷.
viii. Presentation: Removing brackets

Step1: Teacher writes the examples on the chalk board. Remove the following brackets i $6(3a+4b)$
ii $5d(-b-2c)$ iii $-5x(11x-2y)$
solution
i $6(3a+4b) = 6x3a+6x4b= 18a+24b$
ii$5d(-b-2c) = -5bd-10cd$
iii $-5x(11x-2y) = -55x^2+10xy$
Step 2: teacher moves round the class to see students' work
Step 3: teachers gives class exercises i.e
i. expand $(p+q)(r+s)$
ii. $(2-p)(3p+q)$
iii$(7m-5n)(5m+3n)$
iv $(x-4)^2$
solution
i $(p+q)(r+s)$
$p(r+s)+q(r+s)$
$pr + ps + qr + qs$
ii. $(2-p)(3p+q)$
$2(3p+q) - p(3p+q)$
$6p +2q -3p^2 - pq$

iii $(7m-5n)(5m+3n)$
$7m(5m+3n) - 5n(5m+3n)$
$35m^2 + 21mn - 25mn - 15n^2$
iv $(x-4)^2$
$(x-4)(x-4)$
$x(x-4) - 4(x-4) = x^2- 4x - 4x + 16 = x^2 - 8x + 16$
Evaluation:
Teacher evaluates the students by giving them these questions:- Expand the following:
A $(5h+k)^2$
B $(5a-2b)(2a-3b)$
Conclusion: Teacher gives homework NGM BK2 ex7g Q1-10 Pg49.

Stages Four: Give Interim Assessment

Researcher gave:

- formative evaluation by organizing a micro-teaching for the pre-service teachers,
- assessment using the practical teaching assessment rating scale in appendix vii and
- analysis of the pre-service teachers' data again

Five: Re-teaching

Provide remediation until 80% of the pre-service teachers learn before moving to the next topic.

Stage Six: Evaluate and analyse data as feedback

This is the final assessment for a topic. The topic is assessed, graded and scored, the scores were analysed to identify areas pre-service teachers are still finding difficult. This is not to be used for the pre-service teachers but can be used as a feedback for those who will be in the class in future.

Research Question 1: What is the Impact of the Treatment (STM+DDI) on pre-Service Teachers' Lesson Preparation?

Table 2 shows the impact of the treatment on Pre-service teachers' lesson preparation.

Table 2 shows the impact of treatment on Pre-service teachers' lesson preparation in mathematics at [$F_{(1,23)} = 26.827, p<.05$]. The result shows that the treatments (DDI) had an impact on Pre-service teachers' lesson preparation (pre-test and post-test scores) in mathematics. The partial Eta squared estimate was 0.54. This implies that 54% of the variance observed in the pre-service teachers' lesson preparation's post test score was due to treatment (DDI).

Covariates appearing in the model are evaluated at the following values: pre-lesson preparation = 4.68. Table 3 shows that the mean score for pre-service Teacher in DDI is at ($\overline{x} = 7.139$) more than control which is ($\overline{x} = 4.867$).

Research Question 2: Is There a Significant Main Effect of Pre-Service Teachers' Gender on Their Lesson Preparation?

Research Question 3: Is There a Significant Interaction Effect of the Treatment and Teachers' Gender on Pre-Service Teachers' Lesson Preparation?

Table 2. Summary of 2x2 Analysis of Covariance (ANCOVA) of pre-test and post-test of Pre-service teachers' lesson preparation in mathematics by treatment and gender

Source	Type III Sum of Squares	df	Mean Square	F	Sig.	Partial Eta Squared
Corrected Model	67.535[a]	4	16.884	32.649	.000	.850
Intercept	17.595	1	17.595	34.025	.000	.597
Pre-service teachers lesson preparation	6.039	1	6.039	11.678	.002	.337
Treatment(STM+DDI)	13.873	1	13.873	26.827	.000	.538
Teachers gender	.214	1	.214	.414	.526	.018
Treatment * teachers gender	2.259	1	2.259	4.368	.048	.160
Error	11.894	23	.517			
Total	1136.000	28				
Corrected Total	79.429	27				
a. R Squared = .850 (Adjusted R Squared = .824)						

Table 3. Estimated Marginal Means and Standard Error: Treatment

Treatment	Mean	Std. Error	95% Confidence Interval	
			Lower Bound	Upper Bound
DDI	7.139	.269	6.582	7.696
Control	4.867	.273	4.301	5.432

Figure 2. Interaction effect of the treatment and teacher gender on pre-service teachers' lesson preparation
Key 1: male
Key 2: female

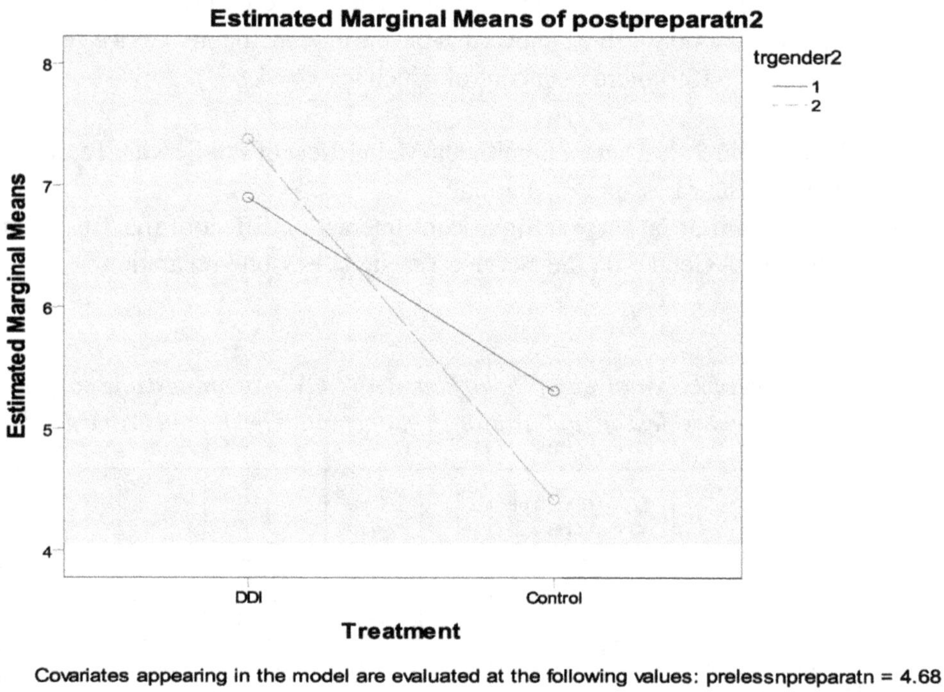

DISCUSSION

Impact of the Treatment (STM + DDI) on Pre-Service Teachers' Lesson Preparation and Pre- Service Teachers' Lesson Delivery

The table shows that the treatment had an impact on pre-service teachers' lesson preparation in mathematics at [$F_{(1,23)}$= 26.827, $p<.05$]. The partial Eta squared estimate was .542. This implies that 54.2% of the variance observed in the pre-service teachers' lesson preparation's post test score was due to treatment (DDI). The finding of this study shows that Data-driven Instructional strategy had an impact on Pre-service teachers' lesson preparation in mathematics as measured by comparison in the means of pre-test score and the post-test score which show statistical difference in the discrepancy of the mean. This study is in agreement with Ankers, Liamas and Tomyoy (2005) investigated the effect of pre-service practice teaching on teachers' career goals, affective measures and classroom teaching and found out that those who had teaching experience had a significantly higher level of confidence in their ability to change student learning in positive ways. The amount of direct supervision they received during student teaching was significantly associated with teachers' desire to remain in teaching.

Main Effect of the Gender on Pre-Service Teachers' Lesson Preparation and Pre-Service Teachers' Lesson Delivery

The study found out that there was no significant main effect of teacher gender on pre-service teachers' lesson preparation at [$F_{(1,23)}$= .414, p>.05]. The result shows that gender had no main effect on Pre-service teachers' lesson preparation (pre-test and post-test scores). The result showed that there was no significant main effect of teachers' gender on pre-service teachers' lesson delivery [$F_{(1,23)}$= .001, p>.05]. The result showed that gender had no main effect on Pre-service teachers' lesson delivery (pre-test and post-test scores). This finding could be as a result of the facts that lesson preparation and lesson delivery skills have a lesser mathematical calculation areas, where females are assumed not to perform well as their male colleagues, at all levels of education (Berube and Glanz, 2008). Another point of view is that, it could have been the effect of the instructional strategy used while making the work comprehensive. The finding that there is no significant difference between the performance of male and female pre-service teachers' lesson preparation and delivery supports many past findings indicating that the gap between the performance of male and female learners in mathematics have been filled (spencer, steel and Quinn, 1999: Austine 2002). Arigbabu and Mji (2004) found out also that score of male pre-service teachers' in academic over three academic sessions was not significantly different to the score of female students in general studies (mathematics). Even though male mean score is higher than that of female but not statistically significant. The finding corroborates Lai (2006) who examined the effect of gender on pre-service teachers'' teaching effectiveness found that there is no difference between both male and female teaching effectiveness.

Interaction Effect of the Treatment (STM + DDI) and Teachers' Gender on Pre-Service Teachers Lesson Preparation and Delivery

Table 2 showed that there was a significant interaction effect of treatment and teachers gender on pre-service teachers' lesson preparation [$F_{(1,23)}$= 4.368, p<.05]. The result showed that the interaction of data-driven instruction is gender sensitive on Pre-service teachers' lesson preparation. This implies that the interaction effect of treatment was significant on Pre-service teachers' lesson preparation in Mathematics and that the treatment was associated with participants' learning outcome. That is, the interaction effect involving the variables influenced by DDI to produce a joint effect. The study aimed at improving Mathematics teachers' Pre-service teachers' lesson preparation and delivery that irrespective of the combination of the variables (treatment, pre-service teachers' gender) to produce a joint effect, the result of the study has shown that DDI group perform better than the control group. Therefore, the findings of this study have provided an empirical support on the effectiveness of DDIS over the conventional method which served as the control group. However, Table 4.3 showed that there was no significant interaction effect of treatment and teachers gender on pre-service teachers' lesson delivery [$F_{(1,23)}$= .257, p>.05]. The result showed that the interaction of data-driven instruction and gender had no main effect on Pre-service teachers' lesson delivery (pre-test and post-test scores). It thus, implies that the effect of treatment on the pre-service teachers' lesson delivery not sensitive to gender. This finding supports the study of Okebukola (1993) as cited by Yinyinola (2008) who believed that all learners can perform equally in a given task, irrespective of sex. Okebukola (1993) as cited by Yinyinola (2008) went further to say that when learners have opportunities to interact among themselves, the teacher and the materials, knowledge and skill are acquired and learning is real for both sexes. Based on this assertion, it is not likely therefore that if Mathematics instructions are presented in a more relevant and interesting way through appropriate access and thought provoking strategies, both male and female learners will perform equally.

CONCLUSION AND RECOMMENDATIONS

This study has been able to establish that Data-driven Instructional Strategy is effective in ameliorating the low performance of learners' achievement in mathematics.

Furthermore, the method of teaching in any teaching-learning environment is governed by the preparation and delivery of basic task to be learned which ultimately has control on learners' attitude. Moreover students' learning outcomes is not sensitive to student gender and also there is no interactive effect that is significant in all cases. The findings of the study have implications for stakeholders in the field of Education:

1. Data-driven instruction is capable of significantly improving teachers' lesson preparation and delivery and also students' learning achievements and attitude towards learning Mathematics. Therefore, it can be said that this instructional strategy is good and can be used to improve students' achievement.
2. This study has shown that there is a need for teachers to ascertain which specific learning strategy is most suitable for various ability levels (both affective and cognitive). Mathematics tutors should have it at the back of their minds that there are different categories of learners in their classes and therefore, use appropriate instructional strategy to carry every learner along.
3. Mathematics teachers should always use students' data to improve learning. In cases where students fail to get a particular concept, teachers should help the students out in breaking the concept through a simpler approach of teaching. They should create more time for the students.
4. School administrators and teachers should plan their scheme of work and lessons in a way that will encourage students' classroom participations.
5. School administrators should implement DDI instructional strategy. There is a need to encourage and persuade secondary school teachers and students to cooperate while it is being used. There is also the need for the school administrators to give students consistent, valid and reliable assessment, in order to generate the data that will drive the teaching learning process. Students' recent performances should always be compared to their past performances to generate improved performance.

REFERENCES

Adewale, J. G., & Ibidiran, O. V. (2013). Construction and validation of affective behaviour of students in Ekiti State West African. *Journal of Education, 33*.

Bambrick-Santoyo, P. (2010). Driven by Data. A practical guide to improve instruction. Jossey Bass.

Bolaji, O. O. (2005). Correlates of Mathematics Achievement and Fear of Success Among Senior Secondary School Students in Kaduna State. *UNAD J. Edu, 3*, 91–101.

Christie, K. (2005, April). Virginia's Excellent Adventure. *Phi Delta Kappan*.

Confrey, J. (2004). Undertaking data analysis of student outcomes as professional development for teachers. *ZDM, 36*(1), 1–9.

Crommey, A. (2000). *Using Student Assessment Data:What Can We Learn From Schools?* Oak Brook, IL: North Central Regional Educational Laboratory.

Federal Republic of Nigeria. (2013). National Policy on Education (6th ed.). Lagos: Nigerian Educational Research and Development Council (NERDC).

Guskey, T. (2005, September). A Historical Perspective on Closing the Achievement Gap. *NASSP Bulletin*.

Hamilton, L. S. (2009). *Making sense of test-based accountability in education*. Santa Monica, CA: Rand.

House, J. D., & Telese, J. A. (2008). Relationships between Student and Instructional Factors and Algebra Achievement of Students in the United States and Japan: An analysis of TIMSS 2003 Educat. *Research Evaluation, 14*(1), 101–112. doi:10.1080/13803610801896679

Popoola, A. A. (2004). *Effects of programmed instructional strategies on Senior Secondary School student learning outcomes in Mathematics in Ekiti State, Nigeria* (An unpublished Ph.D Thesis). University of Ibadan.

Salami, I. A (2014). *Effects of Training Programmes o Pre-service Primary Mathematics Teachers' Planning and Utilization of Activity-Based Lessons in Southwestern States, Nigeria*. Unpublished Ph.D. theses.

Velom, V., & Rohani, F. L. (2010). Interactive Science Homework. An Experiment in Home and School Connections. *National Association of Secondary School Principal Bulleting*, *85*(2), 20–32.

Yara, N. (2009). *Mathematics skills and students achievement*. Oregon Annual Reports.

Yeh, S. (2006, April). High-Stakes Testing: "Can Rapid Assessment Reduce the Pressure?" *Teacher College Record*.

Section 3

Challenges to Mathematics/Science and Modern Globalized Curriculum

Chapter 7
Prospect and Challenges of Mathematics Education in the Modern Globalized Curriculum

Faith Chidinma Nworah
University of Ibadan, Nigeria

Oragade Christy Dolapo
University of Ibadan, Nigeria

ABSTRACT

The pursuit of mathematics for the achievement of prosperity and advancement is very conspicuous in many national developmental plans both in developed and underdeveloped countries. In this chapter, an attempt was made to examine some of the major challenges facing mathematics education in the modern globalized curriculum and its prospect. It was argued that the success of the teaching and learning of mathematics could only be achieved when the subject is given its rightful place. The chapter highlighted the characteristics and usefulness of mathematics, aims and values of mathematics education in the society. Finally, meaningful suggestions were proffered that if adopted will induce a positive change into the teaching and learning of mathematics.

DOI: 10.4018/978-1-5225-6158-3.ch007

INTRODUCTION

School mathematics is traditionally perceived as an elitist subject for more academically oriented learners. In most countries, especially Southern countries, mathematics is a criterion for admission to universities and colleges and it is associated with most of the developments of the technological, information age. Some scholars are concerned that heightened description, prescription, and prediction of the physical and social sciences boosts the utilitarian thrust towards mathematics.

The International Mathematics Study, for example, administers a standard mathematics examination to 13-year-old children from various parts of the world. This examination is aimed at studying teaching and learning, but at the same time it also points to some of the challenges of school mathematics in the era of globalization. In what ways should a uniform mathematics curriculum be taught to 13-year-olds around the world? Critical mathematics scholars voice the need to reflect on the value and the benefits of an excellent performance in an international mathematics examination. Assuming that the standard examination is methodologically accurate and not culturally biased (as mathematics itself is), what does it measure? Despite the possible contributions to mathematics teaching (such as enhancing international collaborations), participation in such international studies are likely to promote international curriculum standards that might negatively influence mathematics teaching.

The last two decades have brought an increased interest in issues of race, gender, and political democracy in mathematics education. Critical theorists criticize school mathematics for its lingering colonial mentality that continues to consider anything different from it as inferior, requiring the work of civilization (Walkerdine, 1990). Given its Eurocentric nature, Wright (1988) maintains, modern mathematics has tended to underplay the contributions of other cultures, and students are rarely informed about such contributions. Mainstream mathematics continues to mistakenly trace its origin solely to Greece, ignoring its historical bases in Egypt, Babylonia, India, and the Middle East, and the parallel mathematical pursuits of the Chinese, the Japanese, Hindus, and Inca-Aztec cultures (P. Davis & Hersh, 1981; Ghever Ghese, 1991). If students want to know the resemblances (probable sources or variations) of Western mathematics to mathematical enterprises in nonwestern traditions and their different tracks of inquiry to what resources can they turn? It seems crucial that school mathematics is encouraged to reflect the historical links of modern mathematics to religion, to socially stratified societies, and to early mathematical traditions. The possible "pre-dation" of Pythagoras' theorem (540 B. C.) in an ancient Indian text, Sulbasutras c. in 800-600 B.C and in a Chinese mathematical source in 500-200 B.C, or the fact that Pascal's triangle may have its origin in China 350 years before Pascal might be of pedagogical significance (Ghever Ghese, 1991; Wright, 1998). At the root of the exclusion of historical development of mathematics from mathematics is the long held belief that mathematics transcends humans and, therefore, is historical.

Pattern of Modern Globalized Mathematics Curriculum

Mathematics can be defined as a group of related sciences, including geometry, calculus and algebra, which is focused on the study of number, space, shape, and quantity, and how they interrelate using a specialized notation. Mathematics has traditionally been described as the science of number and shape. The school emphasis on arithmetic and geometry is deeply rooted in this centuries-old perspective. But as the territory explored by mathematicians has expanded—into group theory and statistics, into optimization and control theory—the historic boundaries of mathematics have all but disappeared. So have the boundaries of its applications: no longer just the language of physics and engineering, mathematics is now an essential tool for banking, manufacturing, social science, and medicine. When viewed in this broader context, we see that mathematics is not just about number and shape but about pattern and order of all sorts. Number and shape—arithmetic and geometry—are but two of many media in which mathematicians work. Active mathematicians seek patterns wherever they arise.

Thanks to computer graphics, much of the mathematician's search for patterns is now guided by what one can really see with the eye, whereas nineteenth-century mathematical giants like Gauss and Poincaré had to depend more on seeing with their mind's eye. "I see" has always had two distinct meanings: to perceive with the eye and to understand with the mind. For centuries the mind has dominated the eye in the hierarchy of mathematical practice; today the balance is being restored as mathematicians find new ways to see patterns, both with the eye and with the mind.

Cyberneticist Norbert Wiener, is one of many exceptional scientists who broke the bonds of tradition to create entirely new domains for mathematicians to explore. Seeing and revealing hidden patterns are what mathematicians do best. Each major discovery opens new areas rich with potential for further exploration. In the last century alone, the number of mathematical disciplines has grown at an exponential rate; examples include the ideas of Georg Cantor on transfinite sets, Sonja Kovalevsky on differential equations, Alan Turing on computability, Emmy Noether on abstract algebra, and, most recently, Benoit Mandelbrot on fractals.

To the public these new domains of mathematics are *terra incognita*. Mathematics, in the common lay view, is a static discipline based on formulas taught in the school subjects of arithmetic, geometry, algebra, and calculus. But outside public view, mathematics continues to grow at a rapid rate, spreading into new fields and spawning new applications. The guide to this growth is not calculation and formulas but an open-ended search for pattern.

Change in the practice of mathematics forces re-examination of mathematics education. Not just computers, but also new applications and new theories have significantly expanded the role of mathematics in science, business, and technology. Students who will live and work using computers as a routine tool need to learn a different mathematics than their ancestors. Standard school practice, rooted in traditions that are several centuries old, simply cannot prepare students adequately for the mathematical needs of the 21st century.

Shortcomings in the present record of mathematical education also provide strong forces for change. Indeed, because new developments build on fundamental principles, it is plausible, as many observers often suggest, that one should focus first on restoring strength to time-honored fundamentals before embarking on reforms based on changes in the contemporary practice of mathematics. Public support for strong basic curricula reinforces the wisdom of the past—that traditional school mathematics, if carefully taught and well learned, provides sound preparation both for the world of work and for advanced study in mathematically based fields.

The key issue for mathematics education is not *whether* to teach fundamentals but *which* fundamentals to teach and *how* to teach them. Changes in the practice of mathematics do alter the balance of priorities among the many topics that are important for numeracy. Changes in society, in technology, in schools—among others—will have great impact on what will be possible in school mathematics in the next century. All of these changes will affect the fundamentals of school mathematics.

To develop effective new mathematics curricula, one must attempt to foresee the mathematical needs of tomorrow's students. It is the present and future practice of mathematics—at work, in science, and in research—that should shape education in mathematics. To prepare effective mathematics curricula for the future, we must look to patterns in the mathematics of today to project, as best we can, just what is and what is not truly fundamental.

Connecting Perspectives of the Roots of Mathematics to the Branches of Mathematics

School tradition has it that arithmetic, measurement, algebra, and a smattering of geometry represent the fundamentals of mathematics. But there is much more to the root system of mathematics—deep ideas that nourish the growing branches of mathematics. One can think of specific mathematical structures e.g numbers, algorithms, ratios, shapes, functions, data, and many others. These diverse perspectives illustrate the complexity of structures that support mathematics. From each perspective one can identify various strands that have within them the power to develop a significant mathematical idea from informal intuitions of early childhood all the way through school and college and on into scientific or mathematical research. A sound education in the mathematical sciences requires encountering virtually all of these very different perspectives and ideas.

Traditional school mathematics picks very few strands (e.g., arithmetic, geometry, algebra) and arranges them horizontally to form the curriculum: first arithmetic, then simple algebra, then geometry, then more algebra, and finally—as if it were the epitome of mathematical knowledge—calculus. This layer-cake approach to mathematics education effectively prevents informal development of intuition along the multiple roots of mathematics. Moreover, it reinforces the tendency to design each course primarily to meet the prerequisites of the next course, making the study of mathematics largely an exercise in delayed gratification. To help students see clearly into their own mathematical futures, we need to construct curricula with greater vertical continuity, to connect the roots of mathematics to the branches of mathematics in the educational experiences of children.

School mathematics is often viewed as a pipeline for human resources that flows from childhood experiences to scientific careers. The layers in the mathematics curriculum correspond to increasingly constricted sections of pipe through which all students must pass if they are to progress in their mathematical and scientific education. Any impediment to learning, of which there are many, restricts the flow in the entire pipeline. Like cholesterol in the blood, poor mathematics instruction can clog the educational arteries of the nation.

In contrast, if mathematics curricula featured multiple parallel strands, each grounded in appropriate childhood experiences, the flow of human resources would more resemble the movement of nutrients in the roots of a mighty tree—or the rushing flow of water from a vast watershed—than the increasingly constricted confines of a narrowing artery or pipeline. Different aspects of mathematical experience will attract children of different interests and talents, each nurtured by challenging ideas that stimulate imagination and promote exploration. The collective effect will be to develop among children diverse mathematical insight in many different roots of mathematics.

For every mathematics prospect there must be some characteristics that will help to know if the mathematics prospect is being aligned and whether it is being achieved or not. Therefore, we will be discussing partially about the characteristics of mathematics.

Characteristics of Modern Mathematics

Mathematics is an essential discipline which forms the basis of all other sciences and deals with the material substance of space and time. Mathematics uses numbers and symbols in the study of measurement, relationships, and properties of quantities and sets. The branches of mathematics include Arithmetic, algebra, geometry, and calculus. Mathematics entails the study of equations, functions, geometric shapes, numbers, equations, and their relationships. Mathematics arises from Man's attempt to summarize the variety of empirical phenomena that he experiences.

But what are the characteristics of mathematics, especially contemporary mathematics?

We shall partially consider five groups of mathematics characteristics:

1. Applicability and Effectiveness.
2. Abstraction and Generality,
3. Simplicity,
4. Logical Derivation, Axiomatic Arrangement,
5. Precision, Correctness, Evolution through Dialectic.

Applicability and Effectiveness of Mathematics

General applicability is a recurring characteristic of mathematics: mathematical truth turns out to be applicable in very distinct areas of application in phenomena from across the universe to across the street. Mathematics is widely useful because the five phenomena that it studies are ubiquitous in nature and in the natural instincts of man to seek explanation, to generalize, and to attempt to improve the organization of his knowledge. As Mathematics has progressively advanced and abstracted its natural concepts, it has increased the host of subjects to which these concepts can be fruitfully applied.

Abstraction and Generality

Abstraction is the generalization of myriad particularities. It is the identification of the essence of the subject, together with a systematic organization around this essence. By appropriate generalizations, many and varied details are organized into a more manageable framework. Put another way, the drive to abstraction is the desire to unify diverse instances under a single conceptual framework. Beginning with the abstraction of the number concept from the specific things being counted, mathematical advancement has repeatedly been achieved through insightful abstraction. These abstractions have simplified its topics, made the otherwise often overwhelming number of details more easily accessible, established foundations for orderly organization, allowed easier penetration of the subject and the development of more powerful methods.

Simplicity (Search for a Single Exposition), Complexity (Dense Exposition)

It is hard to believe that simplicity is a characteristic of mathematics. But, for the practitioner of mathematics, simplicity is a strong part of the culture. The mathematician desires the simplest possible *single* exposition. Through greater abstraction, a single exposition is possible at the price of additional terminology and machinery to allow all of the various particularities to be subsumed into the exposition at the higher level. This is significant: although the mathematician may indeed have found his desired single exposition (for which reason he claims also that simplicity has been achieved), the reader often bears the burden of correctly and conscientiously exploring the quite significant terrain that lies beneath the abstract language of the higher level exposition. Thus, I believe it is the mathematician's desire for a single exposition that leads to the attendant complexity of mathematics, especially in contemporary mathematics.

Logical Derivation, Axiomatic Arrangement

The modern characteristics of logical derivability and axiomatic arrangement are inherited from the ancient Greek tradition of Thales and Pythagoras and are epitomized in the presentation of Geometry by Euclid (The Elements). It has not always been this way. The earliest mathematics was firmly empirical, rooted in man's perception of number (quantity), space (configuration), time, and change (transformation). But by a gradual process of experience, abstraction, and generalization, concepts developed that finally separated mathematics from an empirical science to an abstract science, culminating in the axiomatic science that it is today. It is this evolution from empirical science to axiomatic science that has established derivability as the basis for mathematics.

Precision, Correctness, Evolution Through Dialectic

Mathematical definitions, mathematical notions of correctness, the search for First Principles (Foundations) in Mathematics and the elaboration of areas within Mathematics have all proceeded in a dialectic fashion, alternating between periods of philosophical/foundational contentment coupled with active productive work on the one hand, and the discovery of paradoxes coupled with periods of critical review, reform, and revision on the other. This dialectical process through its history has progressively raised the level of rigor of the Mathematics of each era.

Aims and Values of Mathematics Education in the Society

There is no doubt that mathematics has extensive application in life and related fields. According to observation "mathematics is a gateway to science". Mathematics is a language. It is the language without which science, commerce, industry the internet and the entire global economic infrastructure is struck dump. It is the only "truly" universal language, and it is an essential part of our personal and working life. Mathematics is not only a language and a subject, it is also critical in fostering logical, rigorous thinking, as such, its influence is immense. Everyone recognizes that it is vital to be able to read and write basic language (English). In mathematics, the equivalence of basic reading and writing is numeracy; numbers are the sound, syllables and word of the language of mathematics.

Mathematics is more than just the science of numbers taught by teachers in schools and either enjoyed or feared by many students. It played a significant role in the live of individual, the world and society as a whole. Mathematics is an essential discipline recognized worldwide and it needs to be augmented in education to equip student with skills necessary for achieving higher education or career aspiration, and for attaining personal fulfillment. Since mathematics encompasses all aspects of human life Journal it is unquestionably important in education to help students and all people from all walks of life to perform daily their tasks efficiently and become productive, well informed, functional, independent individuals and members of a society, where mathematics is the fundamental component. Mathematicians create so called models of the world, and study them. This applies even to the simplest mathematics, after the age of five and six we do not study addition by actually combining groups of object and counting them. Instead we use an abstract mathematics contraction or model known as the positive integers i.e (the numbers 3, 4, 5, 6, and so on). Once again, Mathematicians study model so as to idealize world of intangible things that we do not come across in everyday life. For example, talking of a perfect circle, this does not concern tangible worldly thing like chairs, table and so on.

The study of mathematics can satisfy a wide range of interests and abilities, it developed the imagination; it trains in clear and logical thought. It helps in handling varieties of difficult ideas and unsolved problems, because it deals with the question arising from complicated structure, and with a continuous drive to simplification: they are able to finding the right concept and method to make difficult thing easy to explain and understood. In so doing, languages, formulas and insight are developed which may then be applied to make a crucial contribution to our understanding and appreciation of the world around us, and thereby chart our way in it.

CHALLENGES FACING MATHEMATICS EDUCATION

Some of the major challenges are outlined below:

The Challenge of Teachers

Ukeje (1990) in Enoh (1996) opined that the low quality of the teaching personnel constitutes a serious problem in the quest for social reconstruction. Students cannot acquire the necessary skills that are needed in transforming Nigeria in terms of technology without sound and educated teachers.

Going by the common saying that "you don't give what you don't have"; unqualified teachers lack ability to deliver the contents to those they are teaching beyond what they have already known. If the teacher knows little more than the learners, he may not be able to give concrete interpretation to a concept of which greater understanding can be derived. A teacher who possesses a limited understanding of the subject matter cannot deliver effective teaching as he leaves the gap between the contents and its application which invariably will affect student's knowledge of applicability.

Another challenge posed by the teachers is the fact that Mathematics teachers in Nigeria are grossly inadequate. Alagbe (2012) said that many primary and secondary schools have no choice but to employ teachers who have no special training in mathematics because of shortage of mathematics staff.

The Challenge of Students

Many students exhibit negative attitude toward mathematics as they see no importance of the subject in the application of real life. To these young minds, mathematics still remains a mystery that has no place in reality. Vast majority of Nigerian students fear and dread mathematics, because of it's seemingly abstractness of its concepts. Okafor and Anaduaka (2013) observed that many students do not immediately see the use or applicability of the subject to their lives and to the world of work around them; so they wonder why they should be troubled with the study of the subject. Students in secondary schools view mathematics as requirement in gaining admission into higher institutions of learning. Okafor and Anaduaka (2013) said that it is particularly disappointing to find that mathematics has remained one of the least successful subjects in Nigerian schools despite its importance, and time it receives in an average school system.

The Challenge of Curriculum

Curriculum is an important instrument in educational system. As education is central to the society, so is curriculum which is the heart and life-wire of education Alade (2005) in Azuka, Jekayinfa, Durojaiye, and Okwuoza, (2013). Curriculum provides the content of the subject which propels educational programmes and practice. Therefore where this content is not implemented it constitutes a problem of understanding to the students and defies the attainment of the objectives of the curriculum on the students and on the society. For example right from the time of introduction of modern mathematics in Nigeria, it has developed debates as to its applicability and functionality in Nigeria. They perceived the curriculum as foreign in nature having little or no ability in addressing the adequate needs of Nigerians and the Nigeria system. With this raging war in curriculum contents, there is no doubt that ineffective delivery of mathematics concepts to students will be a standing order in the system. The perceived external curriculum contents coupled with the nature of mathematics and its concepts makes mathematics more dreadful and scarring in the mind of the learners.

The Challenge of Society

The wrong value and attitude towards mathematics among the people in the society signals hatred for mathematics. This trend is imbibed by the younger and upcoming students who also show hatred for the subjects. It is common to have Nigerian adults declare their lack of competence in mathematics publicly without any feeling of shame whatsoever. This kind of demoralizing and derogatory statement and attitude towards mathematics displayed the adult members of the society embolden students to continue in a downward trend in their hatred for the subject and lack of mathematical knowledge; this has been made evident with their show of poor performance in the subject during national examinations. Also the emphasy place on certificate rather content has encourage many students to engage in exam malpractice to secure grades in mathematics exams without having the content.

The Challenge of Funding

The issue of funding by the government and other relevant authorities has been a major challenge to mathematics education. It is also important to note that poor funding of education has also contributed to this problem of mathematics Education in Nigeria. The budgetary allocation to the education sector has been grossly inadequate as compared to funding from both developed and some developing countries. In Nigeria, education sector has not received top-most priority in budgetary allocation as it deserves over the years. All these challenges notwithstanding, mathematics education has a lot of potentials and prospects to offer to our teeming youths.

Prospect of Mathematics Education in Nigeria

Mathematics is an exciting and challenging subject which continues to develop at a rapid rate across many research areas. It has a natural elegance and beauty. Taking a real world problem and creating and applying mathematical models to aid understanding is often hugely satisfying and rewarding. Therefore, for our students to enjoy mathematics at school, we must make sure that there are some elements of "good teaching" of mathematics by listening closely to students' ideas, for example, or being sensitively careful at the interface between mathematical and everyday language —are especially important. They are important in order to recruit students into mathematics, as well as to help them succeed there. Mathematics today is having an enormous impact on science and society. The influence may be silent and appear hidden but has shaped our world in many ways. Mathematical ideas have helped make possible the revolution in electronics, which has transformed the way we think and live today. The information technology (IT) of today has transformed the world into a global village. These advances in science and technology are made possible by the numerous developments in pure mathematics. Mathematics is a universal part of human culture. It is the tool and language of commerce, engineering and other sciences –physics, computing, biology etc. It helps us recognize patterns and to understand the world around us. Mathematics plays a vital, often unseen, role in many aspects of modern life, for example: (1) Space travel, (2) Safeguarding credit card details on the internet, (3) Stemming the spread of epidemics, (4) Predicting stock market prices, (5) Business decision making. E.t.c.

As society becomes more technically dependent, there will be an increasing requirement for people with a high level of mathematical training. Analytical and quantitative skills are sought by a wide range of employers. A degree in mathematics provides you with a broad range of skills in problem solving, logical reasoning and flexible thinking. This leads to careers that are exciting, challenging and diverse in nature. Whatever the career plans of our students may be, or if they have no plans at present, a degree in mathematics provides them with particularly good job prospects. To this end, students need to be encouraged to see and know the importance of mathematics in our national development. In view of this fact, quality teaching personnel should be employed to teach the subject. The government policy on recruitment that has been widely criticized and described as abysmal and politically motivated should be dropped while qualified and competent hands should be engaged to salvage the situation.

The prospect of mathematics education will be brighter when the subject teachers' mastery of the subject matter is looked upon. Mathematics teachers should be aware of the fact that "memorization method of teaching mathematics" has no place any longer in the 21st century, this should be replaced with a better method which must be students' centered and friendly; this will unveil the reality of mathematical concepts to the learners. The above will also be a reality when the curriculum planners and developers, in all stages of planning, involve mathematics teachers adequately since they are the ones to implement the contents of the curriculum. This calls for the administrators to regularly organize capacity building workshops and professional development seminars to expose them to techniques of teaching difficult topics/ concepts in mathematics. The general maxim is "if education is expensive, try ignorance" –truly, education is expensive, this is the more reason why budgetary allocation to educational sector should be improved upon by the government, while private sector participation in mathematics education programmes should be encouraged. Notwithstanding, government alone cannot achieve this lofty ideal of effective mathematics education without the full cooperation of the society; the society should lay emphasis on hard work and dignity of labour. "Paper qualification" at all cost, which promotes examination malpractices, should be de-emphasized. Parents will also brighten the prospect of success in mathematics in their wards and society at large, by using positive words to encourage them in the study of the subject, and also by providing them with computers, games, toys that aid mathematics activities.

CONCLUSION

This paper reviewed the Pattern of Modern Globalized Mathematics Curriculum, prospect and challenges of mathematics education. It looked at the Connecting Perspectives of the Roots of Mathematics to the Branches of Mathematics, partially discussed the Characteristics of Modern Mathematics, aims and values of mathematics education in the society and argued that mathematics has great prospect and could play important role in technological development, and prosperity of our educational system, if the teaching of Mathematics is harnessed positively. Now, efforts must also be made by stake holders to put mathematics on a sound footing in order to propel individual nation to the acquisition of a dynamic economy in the committee of nations in the world at large.

Recommendations

This paper recommends the following to achieve the best result in globalized mathematics curriculum:

- Highly qualified and competent mathematics teachers should be employed and current euphemism of "man knows man" should be discouraged as regard recruitment and appointment.
- The Teacher should be actively involved during the curriculum planning
- A cumulative of different countries mathematics curriculum should be collected and reviewed so as to have a unified globalized curriculum
- The curriculum should not be bulky there by creating room for outdoor activities specifically to the learners interest
- Teacher training institutions should put up programmes and incentives to encourage many students to read and study mathematics.
- Regular capacity-building workshops and professional development programmes for mathematics teachers should be organized.
- Wages and salaries of mathematics teachers should be improved upon; this becomes necessary because of the tasking and demanding nature of teaching of mathematics.
- The UNESCO declaration of twenty-six percent (26%) budgetary allocation to the education sector and invariably to mathematics education, should be given accelerated implementation by the government.

REFERENCES

Alade, I. A. (2005). *Rudiments of curriculum studies*. Ibadan: Suflo Publications.

Azuka, B. F., Jekayinfa, O., Durojaiye, D., & Okwuoza, S. O. (2013). Difficult levels of topics in the new senior secondary school mathematics curriculum as perceived by mathematics teachers of federal unity schools in Nigeria. Journals of Educational Practice, 4(17).

Ebrahim, A. (2010). *Characteristics of Modern Mathematics*. Mathematical Science & Technologies.

Etuk, E. D., & Bello, D. O. (n.d.). Challenges and prospects of mathematics Education in Nigeria. *Journal of Assertiveness*. Retrieved from: http://www.globalacademicgroup.com/journals/assertiveness/Emmanuel%20Dan.pdf

Namukasa, I. (n.d.). School Mathematics in the Era of Globalization. *University of Alberta*.

Okafor, C. F., & Anaduaka, U. S. (2015). *Nigerian School Children and Mathematics Phobia: How Mathematics Teachers can help*. Journal of Science and Education Publishing.

Steen. (1998). *The Future of Mathematics Education*. Association for Supervision and Curriculum Development.

Ukeje, B. O. (1990). *Teaching Mathematics in Secondary School*. Anambra Anachuna Educational Books.

Chapter 8
The Challenges and Prospects of Globalized Science Curriculum

Ayodeji Temitope Ojo
University of Ibadan, Nigeria

ABSTRACT

The processes of globalization are leading to widespread changes that are impacting on education worldwide. It has affected education profoundly and in a range of ways. However, for such hyperglobalists, there will be an increasing convergence of educational policies and practices worldwide. Global policies are mediated at the national level through differing cultural and historical traditions and thus produce different national policies in response to the same global pressures. Moreover, the implementation of such national policies in schools has the further potential for mediation according to different cultural traditions both between different countries and within a single country. Hence, the need to discuss the challenges and prospects of globalized science curriculum in Nigeria.

DOI: 10.4018/978-1-5225-6158-3.ch008

INTRODUCTION

The need for explicit teaching has sharpened efforts to understand what knowledge and skills learners are needed for them to engage effectively in learning of science in the science classroom. Today, every sphere of life has been revolutionised by science. It has given us many machines that have made our lives very comfortable. It has become a source of economic might and power. It has provided mankind with remarkable insights into the world we live in. Due to the above reasons, many countries have encouraged the study of science right from a very tender age. In addition, More and more scholars from various fields are recognising the importance and impact of science education, as well as the current and emerging challenges and opportunities to science education (McFarlane, 2013). In his book, "Looking to the Future: Building a Curriculum for Social Activism", Hodson (2011) focuses on Science Education in building intellectual activism for better understanding of, and increase in scientific literary, improved methods in science education, and the development of a more realistic and action-oriented approach to science education.

Hodson (2011) impresses his views to educators, policymakers and administrators, the need to re-engage and redesign curriculum to make science education more formidable and useful as competitive value and solution to our myriad problems. He suggested that science education in the 21st century must focus on developing strategies and solutions to our common problems, and in doing so, must consider the importance of approaches built around collaboration and participatory pedagogy. We are living in a global society where diversity impacts on science call for a variety of perspectives and appreciation of the different learning needs and methods that students and citizens use to understand science even at the most basic level as it unravels in nature. There emerges a necessity to change our understanding of the approach to science literacy as educators, as we recognize that the platform for applying its body of knowledge has changed and is constantly changing (McFarlane, 2013), thus the need for globalized science education

Globalization has become a very prominent process, which has impacted automatically every step of our lives. This is such an encompassing change that has steadily and deeply influenced many aspects of life and our educational institutions also seem to be affected. No area can be seen untouched by this process of Globalization (McFarlane, 2013). According to Vulliamy (2010) it processes have profoundly affects our education across the world. The basic assumption underlying the relationship between Globalization and educational changes is that knowledge is considered an important driving force for Globalization and therefore Globalization should have profound impact on knowledge and education (Carnoy 1999). And the effect is that globalization has changed the system of higher education in a very lucrative market offer nationally and internationally (Anbalagan, 2011). According to Umeasiegbu and Esomonu (2012) global education is concerned with the extent global development issues are integrated into and within the curriculum. It enables the integration and infusion into curriculum of some important world issues, like science and technology, education for sustainability, intercultural education, education for peace and conflict prevention among others. The global education practices present learning as participatory decision-making process which fosters mutual knowledge and collective self-awareness. In this regard, education, knowledge and skill acquisition take the form of global cooperative actions, collaborative reactions and dynamic multidimensional interaction that foster unity among nations. Global education looks at the ways in which everyday life and experiences are affected by the wider world. It stimulates curiosity, motivates learning and contributes to the raising of standards that enable people to take responsibility for their own lives and the world we live in and become active citizens (Umeasiegbu & Esomonu, 2012).

The forces of Globalization have introduced new discourses into curriculum planning. Curriculum in science education is undergoing severe changes that are predominantly influenced by global perspectives and market orientations. This paper identifies 'Globalization and science globalized curriculum, it challenges and prospects in science curriculum.

GLOBALIZATION AND SCIENCE GLOBALISED CURRICULUM

Some disciplines are much more aligned with global perspectives than others. Altbach (2003) is of the view that disciplines and fields vary in terms of how globally similar they have become. Such fields as sciences (information technology and biotechnology among others) are almost entirely dominated by the major academic centres (Altbach, 2003). Various views are expressed in the literature concerning the impact of contemporary Globalization on the structure and processes of science education worldwide. Carnoy (1999) suggested that the direct impact of Globalization on both curriculum and pedagogy at the school classroom level in most countries had to date been minimal. There is little evidence that such an assessment would be any different over a decade later. Thus, for example, whilst there have been some attempts in western industrialised countries to inject more global awareness into the content of school curricula, these have generally remained very low status add-ons.

One of the important implications of Globalization to science curriculum is that there is a growing global perspective (globally homogenous) in science curriculum. The national and local identities are increasingly diluted and compromised in the construction of a globally relevant curriculum. The drive for global perspectives in curriculum to some extent influences curriculum designers to adapt globally homongenous curriculum.

There has, however, been the influence from processes associated with Globalization that is beginning to make a marked impact on teaching styles in schools on the content of the curriculum. This is the widespread introduction of interactive whiteboards and the growing spread of ICT more generally in school classrooms. Research study have shown that nearly most school classrooms had interactive whiteboards in UK (Webb and Vulliamy, 2006). This enabled teachers to access material from the internet in their teaching and pupils to experiment with devising their own PowerPoint presentations.

Camoy (1999) argued that, whilst the direct effects of Globalization on curriculum and pedagogy have been limited, the more general influences of economic re-structuring and associated economic and political ideologies have been great. Thus, for example, changes in the global organisation of work are putting a considerable premium on highly skilled and flexible workers, fuelling the demand both for university education and for lifelong learning. Structural adjustment policies, designed to make countries more attractive to global investors and more competitive economically, involve cuts in the growth of public spending on science education and the promotion instead of user charges for public services and increased privatisation.

Some other important of Globalization to science curriculum are: growing demands for relevance and innovation, emphasis in knowledge application, importance of subject/course specification, competence based learning, regionally and globally comparable degrees, internationalised curriculum, advent of credit-based programmes, introduction of fluid programmes, inter- and trans-disciplinary development, demand for meta-cognitive skills, student centred teaching, self-regulated learning, cultural sensibility, university-industry linkages, market driven degrees, vocationalisation, changing and stratifying the values and importance of disciplines, challenges for universities in resource allocation for different disciplines (David & Wildemeersch 2007).

These above-mentioned implications of Globalization to science curriculum could either have positive or negative outcomes, but the significant educational challenge for science curriculum is to integrate Globalization into the curriculum in a meaningful way so that we can capitalise on – rather than be constrained by – Globalization (Oblinger & Verville,1998).

Challenges of Globalised Science Curriculum

Some of the problems of science globalized curriculum can be traced to the methodological and instructional approaches to science teaching and learning. Science has for too long been taught and learned as a mono-methodological (one single method to answer research question either through qualitative or quantitative) branch of knowledge. Only decades after personal struggles with science literacy are some teachers and experts of science education recognising this as a fundamental problem. This mono-methodological approach emphasised objectivity and empiricism and subjectivity and creativity were penalized in students as teacher-centered approaches to science discredited and deemphasised the kind of participatory pedagogy proposed by Barrett (2008) and the creative-arts-oriented (Barraza, 1999) and action-oriented (Hodson 2011) approaches. Thus, students' creative and self-devised ways of learning science through daily experiences and associations were not exactly appreciated because the "rigor" of science precluded all other matters in the teacher- centered science classroom right up until the 1990s. As Barrett (1999) indicates, participatory approaches to science instruction are not to be perceived as disruptive. However, teachers bent on instructor-dominated or lecture-method approaches tend to view certain levels of participation in science learning from pupils as interference. This attitude needs to change through the practice of embracing more student-friendly and student-centered approaches in the science learning or science classroom. Thus, teacher development becomes a critical factor in the paradigm shift being promoted in science education (Bell & Gilbert, 1996).

The need to understand the importance and relevance of science as an applied and rational field of study immediately accessible in our most immediate and daily environments should have been a major factor prelude to the introduction of science and its methods and body of knowledge in classrooms. However, many teachers continue to make the error of focusing from the beginning on theories and formulas without establishing the raison d'etre of scientific literacy and knowledge to individual personal growth and development and to society. Bartholomew, Osborne and Ratcliffe (Bartholomew, Osborne & Ratcliffe (2004) argue that: learners should appreciate that science is an activity that involves creativity and imagination as much as many other human activities, and that some scientific ideas are enormous intellectual achievements. Scientists, as much as any other profession, are passionate and involved humans whose work relies on inspiration and imagination.

This need for creativity cannot be overemphasized, especially as science education increasingly becomes a competitive factor among nations in the education arms race and economics arms race of the 21st century. Bartholomew *et al* (2004) certainly understand the role of human passion, inspiration and imagination as part of teaching and learning whether it is in science or art. Humans are social creatures and we are socialised into our roles through culture and social interactions from birth to death. Thus, it is only natural and even inherent that our passions and imaginations become the ultimate tools and sources of learning in all we pursue, whether science or art.

Science education is faced with the reality of providing the requisite leadership in ICT via e-learning and global education at all levels of education system. Science education should demonstrate the use of computer in data processing and decision making. It should revolutionised the overall technological output through the system handling and operations of computer for speed, accuracy and efficiency. Science education is therefore challenged with the responsibility of bridging the digital gaps among the citizenry.

According to McFarlane (2013) one of the major challenges of globalised science curriculum in the 21st century is to re-orient the teaching-learning practices to focus on science as "an art" of acquiring scientific knowledge rather than as a rigid and cold method of knowing about our environment, selves and the processes operating at various levels to constitute life processes and affect our survival and ability to use and tame nature and its attributes. Science has for too long been taught as a form of dead poetry. In earlier years of the 18th and 19th centuries, science teachers and learners used their creativity, passion and desire to know and understand as the major factors in fueling scientific rationale (McFarlane, 2013).

The nature of science (NOS) in McFarlane (2013) was not seen as so separate or contrary to the nature of art or literature, but rather a complementary body of knowledge wherein profusive learning was both appreciated and valued. Thus, science and scientific learning fell within the scope of integrative teaching and knowledge acquisition. As technology and mechanization and automation became dominant processes across industries however, the rigorous quantitative indices relevant to performance driven financial and economic units brought about a decline in the diversity of science methods in education. Thus, the globalization and commodification of science (Baskaran & Boden, 2006) has been a significant factor in the changed treatment of science as part of an integrative whole in curricular knowledge requirements in schools and society. Science does not need to respond to Globalization as a laboratory problem, but as a new frontier or platform consisting of diverse elements and particles (McFarlane, 2013).

Prospects of Globalised Science Curriculum in Nigeria

Nigeria as a country still has opportunities to improve in the areas of global education and e-learning. The improvement has to start with innovations in science education and other science based courses. It will require attitudinal changes and discipline without which many breakthroughs in the world would not have been possible. Some of the prospects are as articulated below:

- Development and implementation of curricula that emphasize scientific reasoning, inquiry-based learning and current scientific discovery and understanding of the physical world (Umeasiegbu & Esomonu, 2012).
- In teaching science and physical education, participatory strategies and methods should be used so that students learn by taking responsibilities that cannot be left only to teachers, school authorities, government, and other decision makers (Umeasiegbu & Esomonu, 2012).
- Preparing and nurturing well-educated teachers by recruiting, retaining through reward and valuing good science teachers. These teachers should be offered lifelong professional developments opportunities to improve their content area and pedagogical skills (Umeasiegbu & Esomonu, 2012).
- It has become very important for success in one's career to engage in lifelong learning. The knowledge based economy demands meta-cognitive, higher knowledge and a highly skilled workforce or knowledge professionals. Globalization demands the expansion and massification of science curriculum, as there is an increasing demand for trained citizens for the knowledge-related economy.
- The emphasis on performance, quality, excellence, relevance, competence and innovation should be reflected in science curriculum, which have been promoted by Globalization forces and agents.
- Refresher courses and seminars should be organized for teachers to enable them upgrade themselves in recent and contemporary discoveries in science.
- Information communication technology is of necessity and should be built into curriculum in every subject and course. Relating the industrial output with the laboratory concepts and practice for adoptability in global labour market (Umeasiegbu & Esomonu, 2012).

Science creates and destroys and its creative powers are needed to respond to those who would engage its body of knowledge to destroy what we know as life and social order. From natural disasters to terrorism and medicine, science is one of the most useful and important disciplines in changing our world and the way we live, learn, play and think.

CONCLUSION

The processes of Globalization are leading to widespread changes that are impacting on education world-wide. While nation-states have always felt the need to reform their educational systems in response to international trends and modernizing ideas, Globalization represents a 'new and distinct shift in the relationship between state and supranational forces, and it has affected education profoundly and in a range of ways. However, for such hyperglobalists, there will be an increasing convergence of educational policies and practices worldwide. Global policies are mediated at the national level through differing cultural and historical traditions and thus produce different national policies in response to the same global pressures. Moreover, the implementation of such national policies in schools has the further potential for mediation according to different cultural traditions both between different countries and within a single country.

REFERENCES

Altbach, P. G. (2003). Globalization and the Universities Myth and Realities in an Unequal World. *Journal of Educational Planning and Administration*, *17*(2), 227–247.

Barraza, L. (1999). Children's Drawings About the Environment. *Environmental Education Research*, *5*(1), 49–66. doi:10.1080/1350462990050103

Barrett, M. J. (2008). Participatory pedagogy in environmental education: Reproduction or disruption? In A. Reid, B. B. Jensen, J. Nikel, & V. Simovsla (Eds.), *Participation and learning: Perspective on education and the environment and health* (pp. 212–224). New York: Springer. doi:10.1007/978-1-4020-6416-6_13

Bartholomew, H., Osborne, J., & Ratcliffe, M. (2004). Teaching students? ideas-about-science?: Five dimensions of effective practice. *Science Education*, *88*(5), 655–682. doi:10.1002ce.10136

Baskaran, A., & Boden, R. (2006). Globalization and the commodification of science. In M. Muchie & X. Li (Eds.), *Globalization, inequality and the commodification of life and well-being* (pp. 42–72). London: Adonis & Abbey.

Bell, B., & Gilbert, J. (1996). *Teacher development: A model from science education*. London: Falmer Press.

Carnoy, M. (1999). Globalization and education reform: what planners need to know. UNESCO/IIEP.

David, S. A., & Wildemeersch, D. (2007). *Globalization and Curriculum Restructuring in Higher Education: Comparing the impacts on Higher Education Curriculum Planning in the States Kerala and Tamil Nadu of India (Draft doctoral proposal)*. Katholieke Universiteit Leuven.

Hodson, D. (2011). *Looking to the Future: Building a Curriculum for Social Activism*. Rotterdam, The Netherlands: Sense Publishers. doi:10.1007/978-94-6091-472-0

Oblinger, D. G., & Verville, A. L. (1998). Global Interdependence. In *What Business wants from Higher Education. American Council of Education, Series on Higher Education*. ORYX Press.

Umeasiegbu, O., & Esomonu, N. (2012). E-Learning in Global Education: Challenges and Prospects for Science and Physical Education In Nigeria. *Journal of Education and Practice*, *3*(14), 44–48.

Vulliamy, G. (2004). The impact of Globalization on qualitative research in comparative and international education. *Compare: A Journal of Comparative Education*, *34*(3), 261–284. doi:10.1080/0305792042000257112

Vulliamy, G., & Webb, R. (2006). Education policy under New Labour. In R. Webb (Ed.), *Changing teaching and learning in the primary school*. Buckingham, UK: Open University Press.

Section 4

Ethnomathematics and Globalized Curriculum

Chapter 9 Ethnomathematics and Modern Globalized Currriculum

Omotayo Akintunde
University of Ibadan, Nigeria

Yetunde O. Akanle
University of Ibadan, Nigeria

Efe O. Ogbebor
University of Ibadan, Nigeria

ABSTRACT

The mathematical knowledge consumed in schools can and does influence culture and communities. There is a close connection between development of culture and idea of mathematics. Cultural thinking, practices, and products are mathematically intertwined. Cultural practices show mathematical thinking and operations and culture are better communicated through them. This goes on to confirm the practical fact that mathematics can be understood in cultural game, analysis of local arts, daily work procedures, and skills. Mathematical idea as a science of logical reasoning is better presented from natural/familiar base of culture of the people or else it will not be understood ultimately. Hence, development and human progress cannot be based on it. Mathematics will be useless to humanity. Thus, students need to develop abilities, such as creativity and a sound set of research habits, as they learn the required mathematics. This study focused on ethnomathematics and modern globalized curriculum.

DOI: 10.4018/978-1-5225-6158-3.ch009

ETHNOMATHEMATICS

The prefix ethno originally refers to races, tribes, or groups of relatives. Correspondingly, ethno- mathematics has been associated with the mathematical practices of particular tribes or indigenous, 'primitive' peoples, as well as those of a nation or human race. Until the early 1980s, the notion 'ethno-mathematics' was reserved for the mathematical practices of 'non-literate' – formerly labeled as 'primitive' – people. Furthermore, culture means peoples' language, place, traditions, and ways of organizing, interpreting, conceptualizing, and giving meaning to physical and social worlds. The systems of indigenous knowledge included, particularly, ways of dealing with space and time and different ways of observing, classifying, ordering, comparing, measuring, quantifying, inferring, inventing, plus coherent systems of explanations of facts and phenomena, based on sophisticated founding myths. These are the basic supporting elements of every cultural system and include mathematical ideas present in all these systems, particularly in the Western cultural system, where they are organized as what is called Mathematics.

Ubiratan D'Ambrosio provided a more extended version of this idea, with the aim of encompassing the whole range of cultures. Thus he presupposes "a broader concept of 'ethno', to include all culturally identifiable groups with their jargons, codes, symbols, myths, and even specific ways of reasoning and inferring". Here the word 'ethno' has ceased to refer to anachronistic concepts such as racial groups, primitive peoples or illiterates, and is instead given a comprehensive meaning, pointing to any group of people who share a cultural identity. D'Ambrosio who became the "intellectual father" of the ethno-mathematics program proposed "a broader concept of 'ethno', to include all culturally identifiable groups with their jargons, codes, symbols, myths, and even specific ways of reasoning and inferring". As a result, today, within the ethno-mathematics discipline, scientists are collecting empirical data about the mathematical practices of culturally differentiated groups, literate or not.

The Brazilian mathematician and educationalist Ubiratan D'Ambrosio was the first, from the late 1980s, to propose a research program for ethno-mathematics, he stated that as individuals become integrated into humankind as solidarity, we will be able to contribute to the preservation of a common inheritance". Many people, authors and others, have been attracted to Western-academic mathematics for its elegant, yet powerful sense of logic, form, as well as its scientific, cultural, and artistic qualities. Thus each culture has investigated mathematics in the light of its own needs and activities but the results of their investigation are in agreement with the results of other cultures, i. e the cultural "products" that are observed are common to all cultures included counting, locating, measuring, designing, playing, and explaining.

WHAT IS ETHNOMATHEMATICS?

In the current description of ethno-mathematics as given by the International Study Group on Ethno-mathematics (ISGE) the narrow (anthropological) meaning of ethno-mathematics as well as its broad (sociocultural) meaning are combined as follows.

[Ethno-mathematics] is sometimes used specifically for small-scale indigenous societies, but in its broadest sense the “ethno” prefix can refer to any group—national societies, labor communities, religious traditions, professional classes, and so on. Mathematical practices include symbolic systems, spatial designs, practical construction techniques, calculation methods, measurement in time and space, specific ways of reasoning and inferring, and other cognitive and material activities which can be translated to formal mathematical representation. Ethno-mathematics is the application of mathematical ideas and practices to problems that confronted people in the past or are encountered in present contemporary culture. Much of what we call modern mathematics came about as diverse cultural groups sought to resolve unique problems such as exploration, colonization, communications, and construction of railroads, census data, space travel, and other problems solving techniques that arose from specific communities. Simply put, culture affects the ways we acquire and use our own mathematical knowledge; it is possible to apply ethno-mathematical strategies in teaching and learning mathematics. These strategies include, but are not limited to: historical development of mathematics in different cultures (e.g. the Mayan numeration system), prominent people in different cultures that use mathematics (e.g. an African-American biologist, an Asian-American athlete).

Mathematical applications can be made in cultural contexts (e.g. using fractions in food recipes from different cultures). Social issues can be addressed via mathematics applications (e.g. use statistics to analyze demographic data. In the above context, there is an urgent need to multi culturalize mathematics and its teaching. Towards this, a growing body of literature on ethno-mathematics has been stimulated by concerns for equity, equality, and excellence in a context of diversity. One primary issue is whether mathematics is external or internal to the person. For example, some believe that mathematics is a human invention; others believe that mathematics is embedded in nature and must be discovered.

An increased cultural, ethnic, and racial diversity provides both an opportunity and challenge to societies and institutions, with many questions related to schooling forming an integral part of this question. In this case, ethno refers to groups that are identified by cultural traditions, codes, symbols, myths and specific ways used to reason and to infer. Mathematics is therefore more than counting, measuring, classifying, inferring, and modeling. When a mathematical system is actively used as a system based on theory that can solve real problems, this can be described as modeling.

Figure 1. Interactive curriculum concept
(D'Ambrosio, 1990, p. 376)

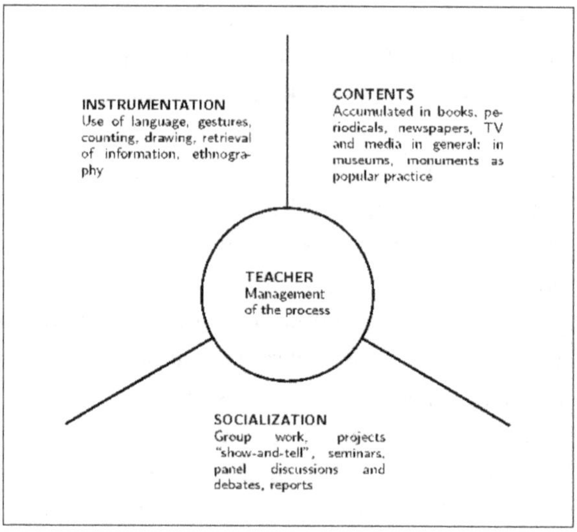

THE ETHNO-MATHEMATICS PROGRAM

D'Ambrosio concept of mathematics which includes a critical, moral, holistic, and global perspective. It is a mathematics curriculum that "walks the mystical way with practical feat, as our philosophical basis to our work in ethno-mathematics and school mathematics. However, other researchers and educators have used a diversity of approaches to demonstrate the dualistic concepts between both approaches in mathematics education (modern western mathematics and ethno mathematics).

Ethno-mathematics forms the intersection set between cultural anthropology and institutional mathematics and utilizes mathematical modeling to solve real-world problems, and translate them into modern mathematical language systems. Mathematical modeling is a tool, which "provides a translation from indigenous knowledge systems to Western mathematics. This perspective is crucial to give minority students a sense of cultural ownership of math, rather than an empty gesture toward inclusiveness and the context allows school mathematics to be seen "as the process of inducting young people into mathematical aspects of their culture".

Ethno-mathematics can reshape our greater cultural identity in a positive way by requiring the inclusion of a greater representation of the true mathematical practices and problems of a student's own community, It helps both educators and students alike to understand the universality of mathematics, in the context of mathematical practices of day-to-day life, preliterate cultures, professional practitioners, workers, and academic or school mathematics. It also develops the study of mathematical ideas and practices of sociocultural groups, but how this is realized in the classroom is still problematic, because much of ethno-mathematical research and investigations identify ethno-mathematical forms of mathematics but do not continue to develop the pedagogical actions for this program from a perspective of using mathematical modeling towards pedagogical action for the ethno-mathematics program. Teaching it is much more than the transference of knowledge because teaching becomes an activity that introduces the creation of knowledge.

ETHNOMATHEMATICS AND STUDENTS

Students need to develop skills in critical thinking and analysis that can best be applied to all areas of life. These skills include vital issues involving health, environment, race, gender, and socio-economic issues. Most researchers agree that ongoing contact of students with diverse and different ways of thinking and practicing mathematics will raise interest in learning the necessary and required content; by having students apply mathematical concepts to future professional contexts.

In an Ethno-mathematics Program, students develop abilities, such as increased creativity, and a sound set of research habits, as they learn the required mathematics. They develop the capacity to create an hypothesis and by developing work in groups, sharing tasks, learning how to appreciate both criticism and alternate opinions, respecting the decisions of others and of the group, and by facilitating student interactions in a globalized society. Students share global and interactive visions, necessary to develop successful mathematical content in our opinion. A major objective of ethno-mathematics as a program can be used to raise student self-confidence.

Teachers and Educators

As performance in mathematics is, on the whole, fairly poor in comparison with other subjects, it is widely argued that something must be fundamentally wrong with its teaching. International studies have indeed shown that, the level of recruitment of math teachers is well below target and that; the enrolment of students in higher math education is also on the decline. This enrolment that has been steadily declining over the past decades could be partly due to a poor public image of mathematics. Ethno- mathematics encourages long-term learning through diversity of experiences.

Students have the right to a multipurpose mathematics; they have the right to be taught thought- provoking mathematics applicable to their lives. Mathematicians, mathematics educators, rural educators can merge their efforts to provide an empirical basis for improving mathematics teaching and learning. This will assist rural teachers to unveil the true mathematics that is significant and constructive to their students. This alternative allows students to learn mathematics content through varied experiences that are related to the cultural, historical, and scientific evolution of mathematics. As a consequence, a broader nature of mathematics was not explored and the use of mathematics was prescribed by these connections bridging schools and student lives. Teachers are characterized as facilitators, advisers, and mediators of the mathematics learning process.

ETHNO-MATHEMATICS AND THE SCHOOL

Just as cultural institutions influences the mode of mathematics teaching, learning, and curriculum. Similarly, the mathematical knowledge consumed in schools can and does influence culture and communities. In both conventional mathematics and the mathematical system of the cultural group are used as activities in ethno-mathematics, it is not just to solve problems, nor that people should understand alternative mathematical systems, but it is also that people should understand more about the role that mathematics play in a societal context. Through modeling it is possible to mix school mathematics and ethno-mathematics. Hence, students practice academic mathematics while learning to model situations which are generated by. This is especially true when they are encouraged to solve problems that are related to their own communities. The selection of mathematical content in Schools should be dependent on situations that are interesting to students themselves, because motivation and creativity are key components in the teaching and learning process in mathematics. In its curriculum model, teachers must learn to be both flexible and knowledgeable in the content because they are the ones who select cases which are related to student cultural background and their environment. School courses and books have presented 'mathematics' as a series of apparently meaningless technical procedures. Such materials are representative of the subject as an account of the name, position, and function of every bone in the human skeleton is representative of the living, thinking, and emotional being called man. Just as a phrase loses meaning or acquires an unintended meaning when removed from its context, so is mathematics when detached from its rich intellectual setting in the culture of our civilization and it is reduced to a series of techniques that has been grossly distorted. A careful look into the history of mathematics reveals that mathematical modeling is the root out of which much of modern mathematics has developed, often from a process of abstraction, which represents the real world. This can also be true in traditional school and academic mathematics.

Students can learn about how different people utilize mathematics within their cultures and about non-Western cultures and subcultures within their neighborhood and all around the world. These activities create an opportunity for teachers to integrate mathematics with other disciplines such as social studies, English, and art, which result in the previously mentioned opportunities for students. Ethno-mathematics activities can be performed with all types of students, from Pre-K to12th grade, and in any type of mathematics classroom, from the academically gifted setting to the standard setting.

Finally, this approach is translated to mathematics education, challenging the instructor to introduce a diversity of mathematical practices in the curriculum.

HISTORICAL DEVELOPMENT OF ETHNOMATHEMATICS

Originated from within cultural anthropology, the focus is on the description of these practices, and of the ideas underlying them (e.g., Ascher, 1998, 2002; Pinxten, 1991; Bazin et al., 2002). A considerable share of case-studies concerns the mathematical practices performed by professional groups such as fishermen, carpenters, carpet weavers, sugarcane farmers, salesmen and vendors (Vithal and Skovsmose, 1997, p. 134). The website of the International Study Group on Ethno mathematics (ISGE) gives a good overview of the ethno mathematical studies from the anthropological point of view. Studies of mathematical practices are listed by ethnicity/geography, e.g., African mathematics, Native American mathematics, Pacific Islander mathematics, African American mathematics, Asian mathematics, Math in European culture, Latino mathematics and Middle Eastern mathematics music from Brazil, card games played in Madagascar, the arithmetic system of the Incas or the Egyptians, basket or carpet weaving, the Mayan calendar, the production of dyes out of natural substances, tread ranking habits or terrace cultivation in China, water collection in the Kalahari desert, the construction of Indian arrows, house construction or sand drawing in Africa, etc.

According to D'Ambrosio (1999) Ethno mathematics is the study of the relationship between mathematics and culture. It is said to be often associated with mathematically oriented practices without written expression. Insightfully, D'Ambrosio (1985) is the founder and most significant theoretician of the ethno mathematics program. The goal of ethno mathematics is to contribute both to clear and logical communication of culture and the understanding of mathematical practices in culture. It mainly leads to an appreciation of the connections between the two.

The samples of some of the definitions of ethno mathematics proposed between 1985-2006 are listed below:

1. The study and presentation of mathematical ideas of traditional people
2. The mathematical idea which is practical among identifiable cultural groups such as tribe, societies, labor group, and children of certain age bracket and professional classes etc.
3. Ascher and Ascher (1994) acknowledged that mathematical ideas exist in all cultures. Ascher (1991) pointed out that this does not imply that, across cultures, mathematical ideas are the same.
4. The study of mathematical ideas of a non-literate culture.

Ethno mathematics emerged as a new conceptual category from the discourse on the interplay among mathematics, education, culture and politics. The development of ethno mathematical ideas can be traceable to the local cultural development of number.

It is discovered that all cultures around the World have basic counting, sorting, ordering, measuring, weighing, and deciphering methods. These have arisen independently in different geographical location around the world. This is a pointer to the fact that mathematical concepts are being discovered rather than created. It is no wonder then that practical concepts as numbers and counting have arisen in all cultures due to needs and challenges in varied human activities. There are some well-known systems of representing numbers in present and previous cultures in different continent of the World. A few are itemized below:

1. **Number Counting:** "Eni bi eni" "eji bi eji" "eta n tagba" etc. "eni" represents one, eji represents two, and "eta" represents three and so forth. This is predominant in western cultures even until today. Other countries around the World have special names for number counting.
2. **Tally System:** This is originated from writing stroke on the wall to represents number. One stroke represents one, two stroke represents two, three strokes represents three, four strokes represents four, four strokes and fifth stroke drawn across the previous marks four strokes represents five then additional one stroke to that gives six and so on and so forth. It teaches addition of number, as the strokes increases the number progresses. It is used across Africa since late Stone Age till today. E.g. Europe, Zimbabwe, Australia, New Zealand, and North America.

3. **Roman Numeral:** It uses few letters of the alphabets to represent numbers in ancient cultural economy up to thousands. It is still in use up till today.
4. **Finger Counting:** This is using fingers of the hands and legs in counting as in tally system. It is the act of counting along one's fingers. In African and most cultures, index finger is used to indicate one, index finger and middle finger indicate two and if together with ring finger indicate three and so on. Finger counting serves as a form of manual communication in market place trading, hand signal during open outcry in auction or floor trading.

ETHNOMATHEMATICS IN LOCAL GAMES IN NIGERIA AND AFRICAN IN GENERAL

"Ayo" Game

"Ayo" game is popularly called "ayo olopon" in the culture of Western Nigeria; other cultures have their special names across African continent. It is a game played by two persons facing each other. It is a wooden board with two rows and six pockets or holes on each row. Each player at a time plays four "ayo" seeds or stones in each of the six pockets or holes. Each of the players takes turn to pick up all the seed or stones in a hole and dropping them one by one round into another holes. Mathematical operations of addition, subtraction and manipulative skills are needed to win the opponent.

Ijakadi

"Ijakadi" is a game of wrestling. Wrestling is a combat game involving two opponents against each other in grapping type techniques, throwing each other; take downs, joint locks, pins down and so on. The aim is for one to gain and maintain a superior position over the other like a total-no-movement-lock-down, put down position and rising over, superior pin down position and so forth.

1. The opponent who is able to:
2. Technically use his intuition faster to give a pin –fall
3. Pre-calculate implication of opponent combative moves
4. Make a counter-balance move to neutralize and out-play his opponent strength or weak areas in skill wins the game.

Figure 2. Ayo competition

Figure 3. Ijakadi competition

It is a highly intellectual activity involving pre-calculation, intuitive, measurement, pre-empty, summary implication fast thinking and fast acting moves to win the opponent. It is common practice among African culture. The games mentioned above are mostly played by adults. Children of nowadays like online games ranging from game consoles, play station, to mortar combat .In the past children congregated outdoor to play their favorite games freely and did not require much equipment or even none. The games are fun to them, build important skills in them, promote social interaction among children, help the children in body co-ordination, primate strategic thinking, and originate basic Mathematics skills in them. Some of other games that are mostly played by children are:

Who Is in the Garden?

This is a game of circular movement exercise of children. It is played by as many children as possible rounded up to form a circle, with one of the children in the middle. One of them will be in a relay run around the rest of the children, while they are singing:

Call: *Who is in the garden?*
Response: *A little fine girl (boy)*
Call: *Can I come and see her (him)?*
Response: *No, no, no, no*
Call: *Follow me.*

The relay runner member of the game then taps the back of the player that he happens to be at his back when the song ends. He becomes the next relay runner. The game continues on and on like that. The circular movement teaches children how to appreciate and understand plain shapes in future.

Figure 4. Children playing Who is in the Garden

Boju-Boju

This is a game played by as many children as possible. One player is the "Oloro" or the catcher, while the rest of the players scatter around in safe place to avoid being caught by the "Oloro" Catcher. The "Oloro" covers his eyes and sings the "Oloro" song thus:

Call: *"Boju boju o"*
Response: *"o"*
Call: *"Oloro n bo"*
Response: *"o"*
Call: *"E Para mo o"*
Response: *"o"*
Call: *"se Ki n si"*
Response: *"si"*
Call: *"si si ki n si eni t'oloro ba mu a paa je"*

Meaning: The "Oloro" catcher is covering his eyes now, but will soon be coming. Should I open my eye, I will devour anyone I see in sight. Shortly at the end of the song, he opens his eyes and pursues all of them. He who the "Oloro" catches becomes the next "Oloro" catcher. It is a high logical game teaching multiple sensing, hides and seeks how to dodge and so forth.

Figure 5. Children playing Boju Boju

Ten-Ten

It is a game mostly played by young girls. It is usually played by two persons, facing each other, clapping and raising their legs simultaneously rhythmically. The rule is not to raise the leg directly opposite the leg which her opponent raised. That is if my opponent raises a left leg, then I should raise a right leg and not the other way round. The player who raises the wrong leg gives the other player a point against herself. It rhythmically movement can get so fast with high speed requiring fast thinking, fast acting, high level of self-articulation and agility.

Suwe

This is a throwing and jumping body exercise game in a rectangular plane shape pattern. It is a mathematical projectile calculation game played by two, three or four persons. First person throw a stone or pebble to the first square pattern in the whole rectangular shape. Then jump with a leg crosses all the square shape without stepping on the border line and the square shape of his stone or pebble. He repeats the same process for all the square shape. At last he backs the rectangular shape and throws the pebble into a square shape which means he has bought a house represented by the square shape into which his pebble has fallen. The second person also steps in to play the game accordingly, and so on for the third and fourth person. The person who breaks any of the rules starts again. The players who eventually buy more houses than the others win the game.

Figure 6. Children playing Ten-Ten

Figure 7. Children playing Suwe

Okoto

This is a game played with a snail shell cuts in form of a cone or a cone shaped soft metal. A player spins the cone shape or snail shell, while it is spinning; he makes a sweeping swipe or cut it with a finger to make the base of the cone lies flat on the ground. The two players do this in turns. The successful one wins and the person who fails offers the back of his hand to the winning player for a forceful strike as a penalty for losing. The fabrication of the cone shaped metal is a natural/cultural intuition that has a mathematical connection.

AMPE

This game can be played by minimum of two people. It is actively played by singing, clapping, jumping, and seems to be a dancing activity. The game is played by splitting the members into groups or forming a semi-circle with the exemption of the leader. The game begins with a jump by the leader, and then every player jumps with the aim of placing a leg forward (left or right) to meet the opposite leg of the opponent first, in order to get a point. Mostly, the person who first gets ten points wins the game. It has survived many generations. It is mostly played in Ghana.

Kudoda

When playing Kudoda game, the players sit forming a ring, around a wooden or metal bowl filled with small stones numbering up to twenty or thirty. The first player pick a stone from the bowl, and project in the air, before it drops down, the player tries to pick as many stones as he/she can. Once all players have had a chance, the stone collected by each player are counted and the one who collects the most stones is the winner. The game has a few variations but all needs at least three players to make it fun. It is mostly played in Zimbabwe.

Skipping

Skipping game involves at least three players and a maximum of four players. Two players stand on each end of a skipping rope and swing it in a circular motion, while the rope itself will form an oval shape mostly while singing a song. A third player jumps in the middle, skipping as the rope turns. The two players on each end starts by swinging the rope low, which is easy for the jumper, then the two players at the ends gradually lift the rope higher from the knee to the thigh, to the waist and on and on… until the player in the middle cannot jump high enough. A fourth player may be introduced to make it more challenging.

Figure 8. Children playing Kadoda

Figure 9. Children skipping

Besides, one single person can engage herself in this game where each competitor skips and every one counts. When the skipper misses the next person starts. The person with the highest number wins the game. It is mostly played by girls across the continent and all over the globe.

Nyama-Nyama-Nyama

This game is played by as many as possible children who stand in a circle, with a leader in the center. Whenever the leader mentions an animal that can be eaten the other children shout "Nyama ! " which literally means meat in Swahili and jump each time an animal is named. The game is played to date across East Africa and is famous among children of all ages.

Mancala Game

This game is regarded as the oldest game around the World. Due to different cultures and customs, different names are given to the mancala game for example it is called kigogo in Kenya, oware in Ghana and other names in other countries. Mancala game is a board game with holes arranged in two or four rows, marbles are used to play the game in recent times in place of stones or seeds. The players target to capture more than the opponent by moving the stones around. It is a game that creates fun but strategically in nature. The game is both enjoy by adult and children across the globe till today.

Figure 10. Adult playing mancala game

ETHNOMATHEMATICS IN LOCAL SKILLS

Pottery

This is the craft of making pot, plates, and cup dish wares out of clay, stones or porcelain. Pottery was developed out of the daily need in day-to-day activities which necessitate fashioning out of different vessel of diverse geometrical shapes for different functional use. Different functional vessels in linear, circular, arcs of two or three-dimensional shapes are formed and improved upon with time.

Textile Weaving

Weaving is a textile art in making clothes. Two distinct set of yarns or thread are interfaced with each other at right angles to form a fabric or cloth. Early weavers in their cultural setting developed it out of calculated interlocking of materials which can give projective sieve for separating materials, protective covering and adornment for men, wall design fabric or interwoven materials to form a container, cage and so forth.

Basket Weaving

Basket weaving is the use of pliable materials like raffia or bamboo to weave two or three-dimensional domestic vessel or artifact like mat, basket, different containers, bird cage and so forth. They are made with equal spacing and crossed pattern according to the functional use for the particular basket, cage, mat or artifact. The dimensional ordered pattern show its ethno mathematical ideas or thinking behind its fashioning.

Woven Designs in Folk Art

Woven designs are used to make cloth, carpet, artistic drawings on walls. It is woven depicting circular, square, cylindrical etc. shapes in a symmetrical arrangement. Local objects and planes are also used in weaving design but design symmetrically to be orderly and good representation of reality. Crowe (2004) shows woven cloth may exhibits one of the seventeen kinds of plane geometry groups. Woven design in most cultures is a local skill in folk art but it has contributed so much to ethno mathematical ideas and practices.

CONCLUSION

In fact, from the foregoing, there is a close connection between development of culture and idea of mathematics. Cultural thinking, practices, and products are mathematically intertwined together. Cultural practices shows mathematical thinking and operations and culture are better communicated through them. This goes on to confirm the practical fact that mathematics can be understood in cultural game, analysis of local arts, daily work procedures and skills.

Mathematical idea as a science of logical reasoning is better presented from natural/ familiar base of culture of the people. Else, it will not be understood ultimately, hence development and human progress cannot be based on it. Mathematics will be useless to humanity.

REFERENCES

Ascher, M. (1991). *Ethnomathematics: A Multicultural View of Mathematical Ideas*. Pacific Grove, CA: Brooks/Cole.

Ascher, M. (1998). *Ethnomathematics. A Multicultural View of Mathematical Ideas*. New York, NY: Hapman and Hall /CRC.

Ascher, M. (2002). *Mathematics Elsewhere. An Exploration of Ideas Across Cultures*. Princeton, NJ: Princeton University Press.

Ascher, M., & Ascher, R. (1994). Ethnomathematics. In Companion Encyclopedia of the History and Philosophy of the Mathematical Sciences (pp. 1545–1554). Routledge.

Bazin, M., & Tamez, M. (2002). *Math and Science Across Cultures. Activities and Investigations from the Exploratorium*. New York, NY: The New Press.

D'Ambrosio, U. (1985). Ethnomathematics and its place in the history and pedagogy of mathematics. *For the Learning of Mathematics*, *5*(1), 44–48.

D'Ambrosio, U. (1990). *Etnomatemática* [Ethnomathematics]. São Paulo: Editora Ática.

D'Ambrosio, U. (1999). Literacy, Matheracy, and Technoracy: A Trivium for Today. *Mathematical Thinking and Learning*, *1*(2), 131–153. doi:10.120715327833mtl0102_3

Pinxten, R. (1991). Some remarks on Navajo geometry and Piagetian genetic theory. *Infancia y Aprendizaje*, *54*(54), 41–52. doi:10.1080/02103702.1991.10822292

Vithal, R., & Skovsmose, O. (1997). The end of innocence: A critique of ethnomathematics. *Educational Studies in Mathematics*, *34*(2), 131–157. doi:10.1023/A:1002971922833

Chapter 10
Ethno-Science and Globalization

Saheed Ayodeji Adejimi
University of Rwanda, Rwanda

ABSTRACT

Globalization accelerates at an extreme rate through development of technologies of communication. However, people living in poor countries' access is very difficult because of the energy and financial demands that technology requires. One consequence of globalization for science and science education is it creates even more complex societies and challenges for indigenous communities. However, increasing local achievement in science and science education is advocated by a number of researchers in order to provide opportunities for people globally. This issue of making local knowledge part of the global brings with it the challenges of politics, history, language, economics, and ethics. The effects of globalization have been far-reaching, while the living standards of the world are still highly uneven. This study focused on ethno-science and globalization.

DOI: 10.4018/978-1-5225-6158-3.ch010

INTRODUCTION

Ethno means to view an event or occurrence from one's cultural perspective. Science on the hand is the fact of knowing something; knowledge or understanding of a truth. Ethnoscience can be said to be the understanding/ knowledge of a particular culture scientifically. Ethnoscience, viewed from a generalized perspective is the study, which approximates or reflects the native's thinking about how their physical world is to be classified, consciously or unconsciously, explicitly or implicitly, within the framework they accordingly act. Abonyi (2000), states that ethnoscience is the knowledge that is indigenous to a particular culture. Ethnoscience according to Atran (1991: 650) a scientific realm which aims to understand how humans—in spite of their fragmented and limited interactions with the world—are developing different forms of knowledge and beliefs. This allows an understanding of how humans vary their knowledge and beliefs within different ecological and historical contexts in order to express the manifold possibilities offered by human cultures.

More concretely, ethnosciences are referring to the set of concepts, prepositions, and theories that are unique to each particular culture group in the world (Meehan, 1980). Ethnosciences are essentially cross-disciplinary, based on increased collaboration between social sciences and humanities (anthropology, sociology, history of science, psychology, philosophy) with natural sciences such as biology, ecology, agronomy, climatology, astronomy, or medicine.

PRINCIPLES OF ETHNOSCIENCE

Etics and Emics

Etic and emic are terms coined by linguistic anthropologist Kenneth Pike, Franklin (1996), which were derived from an analogy with the terms "phonemic" and "phonetic". Etic categories involve a classification according to some external system of analysis considered as appropriate by science. This is the approach of biology where the Linnaean classification system is used to define new species. It assumes that ultimately, there is an objective reality that is seen to be more important than cultural perceptions of it. In contrast, emic categories involve a classification according to the way in which members of a society perceive and classify their own world.

If a folk classification is ever to be fully understood, an ethnoscientific analysis must ultimately reduce to a description in terms approximating culture-free characteristics. Culture-free features of the real world may be called "etics" (Pike 1954). The label may also be applied to features which are not truly culture-free, but which at least have been derived from the examination of more than one culture, or to the sum of all the significant attributes in the folk classifications of all cultures. Emics in contrasts is an attempt "to discover and describe the behavioral system [of a given culture] in its own terms, identifying not only the structural units but also the structural classes to which they belong" (French 1963:398). An emic description should ultimately indicate which etic characters are locally significant. The more we know of the etics of culture, the easier is the task of ethnoscientific analysis.

Domains

One of the most important principles of ethnoscience, and one of those most often overlooked, is the necessity for determining in a nonarbitrary manner the boundaries of the major category or classification system being analyzed, i.e., for discovering how a domain is bounded in the culture being described rather than applying some external, cross-cultural definition of the field, If this is not done, the description of the internal structuring of the domain is likely to be incomplete if not entirely erroneous, and the utility of the analysis for predicting the classificatory placement of new instances will suffer. (Hymes, 1964: 16-18.) Any two cultures differ in the way they classify experience.

Terminological Systems

Research in ethnoscience so far has concentrated on classifications as reflected by native terminology, on "discerning how people construe their world of experience from the way they talk about it" (Frake 1962: 71). The analysis of a culture's terminological systems will not, of course, exhaustively reveal the cognitive world of its members, but it will certainly tap a central portion of it. Culturally significant cognitive features must be communicable between persons in one of the standard symbolic systems of the culture. A major share of these features will undoubtedly be codable in a society's most flexible and productive communication device, its language (Frake 1962: 75; Conklin 1962a, Goodenough 1957, Lounsbury 1963). One task of ethnoscience, in fact, can be viewed as the solution of the old problem of translation.

Paradigms and Componential Analysis

A key concept in ethnoscience is that of the contrast set. This is a class of mutually exclusive segregates which occur in the same culturally relevant environment (setting, context, substitution frame, surroundings, situation, etc.). These segregates "share exclusively at least one defining feature"-i.e., that which characterizes the environment in which they occur (Conklin 1962a: 124; cf. Frake 1962:78-79). The domain of the set is the total range of meanings of its segregates. The notion of contrast is relative to the environment within which it occurs.

A paradigm is a set of segregates which can be partitioned by features of meaning, i.e., a set some members of which share features not shared by other segregates in the same set (Chafe 1962; Conklin 1962a: 132; Goodenough 1956a: 197, 202; Lounsbury 1960: 127-28, 1962). A set of only two segregates can be considered a paradigm, but normally the term is applied to sets of three or more segregates, so that at least some of the sub-sets consist of two or more segregates sharing some feature of meaning. It is important to note that while all contrast sets are paradigmatic, not all paradigmatic sets are complete contrast sets. A paradigmatic set may not be equivalent to its containing contrast set: it is possible to analyze paradigmatically a collection of items which do not exclusively share any feature, which do not exhaust the membership of a class occurring in a single environment (Conklin 1964).

A componential analysis is an analysis of a paradigm in terms of the defining features, the "dimensions of contrast" or "criterial attributes" of the segregates in the set. The aim is to discover the "rule for distinguishing newly encountered specimens of [a] category from contrasting alternatives" (Frake 1962:83). The procedure is to search for the minimum features of meaning which differentiate segregates in the set. Each feature has two or more contrasting values, termed "components." Each segregate is then defined in terms of the presence or irrelevance of each component; i.e., a bundle of components defines the segregate. It is normally assumed that the number of componential dimensions will be smaller than the number of segregates they define. The paradigm may then be viewed as a multidimensional structure, in which the categories are placed according to the componential dimensions.

Taxonomies

Different segregates within a folk classification may be related to each other in various ways: as part to whole, as sequential or developmental stage to stage, as different grades of intensity, etc. (Conklin 1962a: 129, 1962b; Frake 1964). The kind of relationship between segregates which has so far received the most attention is that of inclusion; segregates related in this way form a taxonomy-a folk taxonomy in the case of folk classifications. In a taxonomy, there is a series of hierarchical levels, with each segregate at one level included in (only) one segregate at the next higher level. It is sometimes possible to analyze componentially a contrast set which forms one level of a folk taxonomy, but it is impossible to analyze in this way the whole taxonomy, even though the boundaries of the whole must define a domain: a single contrast set is limited to one taxonomic level (Conklin 1962a: 128, 1964; Frake 1962). A single folk classification may contain sets of segregates interrelated in different ways. Folk taxonomies have hierarchical levels similar to formal biological classifications of kingdom, phylum, class, order, family, genus and species. (Berlin, 1992). In folk taxonomy, the common levels are:

- **Life-Form:** A high level of plants or animals that share some general shape or characteristics in morphology. Examples: tree, vine, bush, fish, snake, bird, mammal.
- **Generic:** The most common basic level. Examples are dog, grass and rice ant. Folk genera often do not correspond to scientific genera but sometimes to Linnaean species, "ant" is a folk genus and belongs to Linnaean family formicidae.
- **Specific:** Usually separated from each other by few characteristics. In some languages, such as Spanish, Bahasa, Indonesia and Malaysia, the generic name comes first, as in a Linnaean name. In English, Filipino, Chinese, Vietnamese and Thai, it is the other way round. Specific names tend to be pneumonic device—like colour, shape and utility—that makes the names easy to remember.

Discovery Procedure

Since the ethnoscientific method aims at discovering culturally relevant discriminations and categorizations, it is essential that the discovery procedures themselves be relevant to the culture under investigation. While arbitrary stimuli-i.e., stimuli foreign to the culture-may yield nonrandom responses, the patterning involved derives from the cognitive system of the bearers of the culture, and the principles of this system are not likely to be made clear by answers to the wrong questions. Regularities will appear if one measures continental European manufactured goods with an American or British yardstick, but measuring with a meter stick will much more readily reveal the principles of the system relevant in European culture.

If an ethnography is to reflect the cognitive system of the bearers of a culture, the validity of the description depends on the discovery procedures. Hypotheses must be checked in the field situation, and revised if they turn out not to fit the field data. Thus it is impossible to make a strictly ethnoscientific analysis of data previously collected, by oneself or by someone else, according to different procedures. Any componential or similar analysis made of such old data must be treated as an inadequately checked hypothesis. Structural restatements of even the best old field data may prove impossible (Sturtevant, 2009).

Globalized Science Education Curriculum

Globalization is one aspect of the larger phenomenon of modernization, which describes societies characterized by progressive growth in the complexity of communications (Luhmann, 1995; Charlton and Andras, 2003). Globalization is a widely-used term describing processes of global (i.e. worldwide) distribution of ideas and goods, most significantly and influentially, regarding scientific, technological and cultural products and developments.

Globalization not only refers to a series of economic changes but technological changes that have changed the way the world works and transfers information (Penn, 2005, p. 4). These changes include: changes in the financial markets, changes in international trade, changes in investment patterns, even changes in the way crimes are carried out. Brown and Lauder (1996) extended this definition of globalization to include "a change in the rules of eligibility, engagement, and wealth creation" (p. 2). The rules of eligibility have changed to either include or exclude certain groups. The included are wealthy, highly mobile, and typically Western with certain skills (Freidman, 2000) while the excluded are incorporated into the global world by the dominance of the global markets. Therefore, the new terms of eligibility due to globalization are creating a new binary view from have/have-not into eligible/ineligible (Carter, 2008). Those that are eligible participate in new modes of wealth creation, changing from industrial work to the ever-changing knowledge that design, produce and market products and services. Specifically, this is having a profound effect on scientific knowledge moving away from fundamental inquiry towards discovery of new products and services (Carter, 2008).

Globalization refers specifically to the increasing dominance of an international network of communications – especially in the economy, but also in social systems such as politics, the mass media, and science and technology. Globalization has helped to significantly reduced the distance between research partners in the contemporary world due to improvement in transportation system, ICT facilities and the rise of English language in mainstream science. Part of this pervasive process is driven by the internationalization ('globalisation') of the higher education markets (Wildavsky, 2010), part by international competition among research universities and academic researchers within the leading science nations (Mohrman, Mab, & Baker, 2008).

Globalization has meant that at the local level, the world's peoples rub more closely together ensuring that diversity, plurality, hybridity and dislocation have become the leitmotifs of the global age (Carter, 2007).

Globalization has enables and facilitates increasing specialization and coordination of functions; for instance the division of labour in a large organization such as a factory or hospital, or the differentiation of social functions in a modern society where major communication systems (such as the economy, law, the mass media and science) operate using functionally distinctive selections, evaluations and rules (Luhmann, 1995, Charlton and Andras, 2003).

EFFECTS OF GLOBALIZATION ON SCIENCE

Today, globalization continues to accelerate at an extreme rate through development of technologies of communication, however people living in poor countries access is very difficult because of the energy and financial demands that technology requires. One consequence of globalization for science and science education is it creates even more complex societies and challenges for indigenous communities. Similarly to colonization globalization changes indigenous knowledge systems, even if that is not the goal (Cassie, 2009). Globalization, despite its inevitable problems, is a generally desirable phenomenon, since it enables increased efficiency, effectiveness and capability of societies and thereby potentially benefits most people most of the time (Luhman, 1997). As globalization continues to increase, it allows for contact between once geographically isolated groups, and traditional knowledge systems are being assimilated and in some cases disappearing all together. For many indigenous peoples, this type of culture is one of colonizing, although due to increased globalization, the means of colonizing is changing.

In a time of globalization in terms of technology and increased worldwide travel where populations migrate, indigenous knowledge is often dismissed as irrelevant and the Internet makes location an intangible concept. However, increasing local achievement in science and science education is advocated by a number of researchers in order to provide opportunities for people globally (McKinley, 2005). This issue of making local knowledge part of the global brings with it the challenges of politics, history, language, economics and ethics.

The effects of globalization have been far-reaching, while the living standards of the world are still highly uneven, 400 million people have moved out of extreme poverty since 1980- more than at any other time in human history (Stewart, 2012).

REFERENCES

Abonyi, S. O. (2000). *Science and Technology Education for Sustainable for Development. A case for Ethnoscience*. Retrieved from https://www.researchgate.net/publication/260988590

Atran, S. (1991). Ethnoscience today. *Social Science Information Sur Les Sciences Sociales*, *30*(4), 595–662. doi:10.1177/053901891030004001

Berlin, B. (1992). *Ethnobiological classifications: Principles of categorization of plants and animals in traditional societies*. Princeton University Press. doi:10.1515/9781400862597

Brown, P., & Lauder, H. (1996). Education, globalization and economic development. *Journal of Education Policy*, *11*(1), 1–25. doi:10.1080/0268093960110101

Carter, L. (2007). Sociocultural Influences on Science Education: Innovation for Contemporary Times. Wiley InterScience. *Science Education.*, 166–181. doi:10.1002ce

Carter, L. (2008). Globalization and science education: The implications of science in the new economy. *Journal of Research in Science Teaching*, *45*(5), 617–633. doi:10.1002/tea.20189

Cassie, Q. (2009). Globalization and Science Education: The Implications for Indigenous Knowledge Systems. *International Education Studies*, *2*(1), 76–88. Retrieved from www.ccsenet.org/journal.html

Charlton, B., & Andras, P. (2003). *The modernization imperative*. Exeter, UK: Imprint Academic.

Conklin, H. C. (1962) Lexicographical treatment of folk taxonomies. International Journal of American Linguistics, 28(2), 119-141.

Frake, C. O. (1962). The ethnographic study of cognitive systems. In Anthropology and Human Behavior. Washington, DC: Anthropological Society.

Franklin, K. J. (1996). *K.L Pike on etic vs emic: A review and interview*. Retrieved from www.sil.org/klp/karlintv.htm

French, D. (1963). The relationship of anthropology to studies in perception and cognition. In S. Koch (Ed.), *Psychology: A Study of a Science* (Vol. 6, pp. 388–428). New York: McGraw-Hill. doi:10.1037/10590-007

Goodenough, W. H. (1957). Cultural anthropology and linguistics. In *Report of the 7th Annual Round Table Meeting on Linguistics and Language Study*. Georgetown University.

Lounsbury, F. G. (1960). Similarity and contiguity relations in language and in culture. In *Report on the 10th Annual Round Table Meeting on Linguistics and Language Studies*. Georgetown University.

Luhmann, N. (1995). *Social systems*. Cambridge, MA: Harvard University Press.

Luhmann, N. (1997). Globalization or world society: How to conceive of modern society. *International Review of Sociology*, *7*(1), 67–80. doi:10.1080/03906701.1997.9971223

McKinley, E. (2005). Locating the global: Culture, language and science education for indigenous students. *International Journal of Science Education*, *27*(2), 227–241. doi:10.1080/0950069042000325861

Meehan, P. (1980). Science, enthnoscience, and agricultural knowledge utilization. In D. M. Warren, D. Brokensha, & O. Werner (Eds.), *Indigenous knowledge systems and development* (pp. 383–391). Lanham, MD: University Press of America.

Mohrman, K., Mab, W., & Baker, D. (2008). The research university in transition: The emerging global model. *Higher Education Policy*, *21*(1), 5–27. doi:10.1057/palgrave.hep.8300175

Penn, H. (2005). *Unequal Childhoods: Young children's lives in poor countries*. London: Routledge. doi:10.4324/9780203465349

Pike, K. (1954). *Emic and etic standpoints for the description of behaviour*. Glendale, IL: Summer Institute of Linguistics.

Stewart, V. (2012). *World-Class Education. Learning from international of world of experience and innovation*. Retrieved from www.ascd.org

Sturtevant, C. W. (2009). Anthropological Approaches. Studies in Ethnoscience. *American Anthropologist*, *66*(3). doi:10.1525/aa.1964.66.3.02a00850

Wildavsky, B. (2010). *The great brain race: How global universities are reshaping the world*. Princeton, NJ: Princeton University Press.

Compilation of References

Abonyi, S. O. (2000). *Science and Technology Education for Sustainable for Development. A case for Ethnoscience*. Retrieved from https://www.researchgate.net/publication/260988590

Adebagbo, A. O. (2004). *Effect of Games on Primary Pupils' Learning Outcomes in Developmentof Multiplication Skills* (Unpublished M.Ed Dissertation). Department of Teacher Education, University of Ibadan, Ibadan, Nigeria.

Adegboye, A. O. (1999). The assessment of laboratory approach in the teaching of mathematics. *Nigeria Education Digest, 1&2*, 149-162.

Adewale, J. G., & Ibidiran, O. V. (2013). Construction and validation of affective behaviour of students in Ekiti State West African. *Journal of Education, 33*.

Adeyemo, D. A., & Adetona, M.O. (2007). A path analytic study of the personality factors affecting students' learning outcomes in Mathematics. *European Journal of Scientific Research, 18*(1).

Ajobiewe, T. (2016). *Computer appreciation training programme. Federal College of Education (Special)*. Ibadan Binta Image Resources Nig. Ltd.

Akinoso, S. O. (2012). *Effects of Concrete-Representational-Abstract and Explicit Instructional Strategies On Senior Secondary School Students' Achievement In And Attitude To Mathematics*. A Ph.D Post-Field Report Presented At The Joint Staff/ Higher Degree Students Seminar, Department Of Teacher Education, Faculty Of Education, University Of Ibadan, Ibadan, Nigeria.

Akinsola, M. K. (1999). Effect of instruction on students' performance on knowledge comprehension and application tasks in mathematics. *African Journal of Education Research, 5*(1), 94–100.

Akinsola, O. S. (2001). The extent of dominant factors of girls and women's low participation in Mathematics: Mathematics educators' views. *Proceedings of the 42nd Annual Conference of the Science Teachers Association of Nigerian Mathematics Panel, 295-298.*

Akpan, A. A. (1987). *Correlates of Mathematical Problem-Solving Ability among Secondary School Students* (Unpublished Ph.D. Thesis). University of Ibadan, Ibadan, Nigeria.

Alade, I. A. (2005). *Rudiments of curriculum studies*. Ibadan: Suflo Publications.

Alele, W. G. (1988). Keynote Address Delivered at the Silver Jubilee meeting of mathematics Association of Nigeria (MAN) Sept. 1982. *Abacus, 18*(1), 15–19.

Ale, S. O. (1989). School Mathematics in the 1990's some Major Problems for Developing Countries. *International Journal of Mathematical Education in Science and Technology, 20*(5), 655–659. doi:10.1080/0020739890200502

Altbach, P. G. (2003). Globalization and the Universities Myth and Realities in an Unequal World. *Journal of Educational Planning and Administration, 17*(2), 227–247.

Amobi, F. A. (2006). Beyond the call: Preserving reflection in the preparation of 'high qualified' teacher. *Teacher Education Quarterly, 33*(2), 23–33.

Anderson, S. E. (1990). Worldmath curriculum: Fighting euro-centrism in mathematics. *The Journal of Negro Education, 59*(3), 348–359. doi:10.2307/2295569

Animashahun, I. A. (2002). The Effects of Simulation-Games on Students' Achievement in and Attitude to Mathematics (Unpublished M.Ed Project). University of Ibadan.

Annetta, L., Mangrum, J., Holmes, S., Collazo, K., & Cheng, M.-T. (2009). Bridging reality to virtual reality: Investigating gender effect and student engagement on learning through video game play in an elementary school classroom. *International Journal of Science Education, 31*(8), 1091–1113. doi:10.1080/09500690801968656

Apple, M. W. (1992). Do the standards go far enough? Power, policy and practice in mathematics education. *Journal for Research in Mathematics Education, 23*(5), 412–431. doi:10.2307/749562

Aremu, S. A. (1998). *Effects of card and Geoboard game-based Instructional Strategies on Primary School Pupils' Achievement in Practical Geometry* (Unpublished Ph.D Thesis). Department of Teacher Education, University of Ibadan.

Armory, A., Naicker, K., Vincent, J., & Adams, C. (1999). The use of computer games as an educational tool: Identification of appropriate game types and game elements. *British Journal of Educational Technology*, *30*(4), 311–321. doi:10.1111/1467-8535.00121

Ascher, M., & Ascher, R. (1994). Ethnomathematics. In Companion Encyclopedia of the History and Philosophy of the Mathematical Sciences (pp. 1545–1554). Routledge.

Ascher, M. (1991). *Ethnomathematics: A multicultural view of mathematical ideas*. Pacific Grove, CA: Brooks/Cole Publishing Company.

Ascher, M. (1991). *Ethnomathematics: A Multicultural View of Mathematical Ideas*. Pacific Grove, CA: Brooks/Cole.

Ascher, M. (1998). *Ethnomathematics. A Multicultural View of Mathematical Ideas*. New York, NY: Hapman and Hall /CRC.

Ascher, M. (2002). *Mathematics Elsewhere. An Exploration of Ideas Across Cultures*. Princeton, NJ: Princeton University Press.

Atran, S. (1991). Ethnoscience today. *Social Science Information Sur Les Sciences Sociales*, *30*(4), 595–662. doi:10.1177/053901891030004001

Atweh, B., & Clarkson, P. (2005). *More perspectives on the impact of globalisation on mathematics education in higher education*. Retrieved on 14th November, 2011 from www.merga.net.au

Atweh, B., Clarkson, P., & Nebres, B. (2003). Mathematics education in international and global contexts. In *Second international handbook of mathematics education* (pp. 185–229). Springer Netherlands. doi:10.1007/978-94-010-0273-8_7

Ayotola, A., & Adedeji, T. (2009). The relationship between gender, age, mental ability, anxiety, mathematics self-efficacy and achievement in mathematics. *Cypriot Journal of Educational Sciences*, *4*(2), 113–124.

Azuakola, S. (2012). *Naija.com*. Retrieved from ynaija.com/better-but-far-from-good-waec-releases-mayjune-2012-ssce-results/

Azuka, B. F. (2001). *The use of Mathematical Games in Nigerian Secondary Schools*. Paper presented at Mathematical Association Nigeria Conference, Katsina.

Azuka, B. F., Jekayinfa, O., Durojaiye, D., & Okwuoza, S. O. (2013). Difficult levels of topics in the new senior secondary school mathematics curriculum as perceived by mathematics teachers of federal unity schools in Nigeria. Journals of Educational Practice, 4(17).

Bambrick-Santoyo, P. (2010). Driven by Data. A practical guide to improve instruction. Jossey Bass.

Banks, J. A., & Banks, P. (2013). *Multicultural Education, Multicultural Education: Characteristics and Goals, Culture, Teaching and Learning*. John Wiley & Sons.

Barraza, L. (1999). Children's Drawings About the Environment. *Environmental Education Research*, *5*(1), 49–66. doi:10.1080/1350462990050103

Barrett, M. J. (2008). Participatory pedagogy in environmental education: Reproduction or disruption? In A. Reid, B. B. Jensen, J. Nikel, & V. Simovsla (Eds.), *Participation and learning: Perspective on education and the environment and health* (pp. 212–224). New York: Springer. doi:10.1007/978-1-4020-6416-6_13

Bartholomew, H., Osborne, J., & Ratcliffe, M. (2004). Teaching students? ideas-about-science?: Five dimensions of effective practice. *Science Education*, *88*(5), 655–682. doi:10.1002ce.10136

Baskaran, A., & Boden, R. (2006). Globalization and the commodification of science. In M. Muchie & X. Li (Eds.), *Globalization, inequality and the commodification of life and well-being* (pp. 42–72). London: Adonis & Abbey.

Bazin, M., & Tamez, M. (2002). *Math and Science Across Cultures. Activities and Investigations from the Exploratorium*. New York, NY: The New Press.

Bell, B., & Gilbert, J. (1996). *Teacher development: A model from science education*. London: Falmer Press.

Berlin, B. (1992). *Ethnobiological classifications: Principles of categorization of plants and animals in traditional societies*. Princeton University Press. doi:10.1515/9781400862597

Bishop, A. J. (1988). *Mathematical enculturation: A cultural perspective on mathematics education*. Dordrecht: Kluwer Academic Publishers. doi:10.1007/978-94-009-2657-8

Blumberg, F., & Sokol, L. M. (2004). Boys' and Girls' Use of Cognitive Strategy When learning to Play Video Games. *The Journal of General Psychology*, *131*(2), 151–158. doi:10.3200/GENP.131.2.151-158 PMID:15088867

Bolaji, O. O. (2005). Correlates of Mathematics Achievement and Fear of Success Among Senior Secondary School Students in Kaduna State. *UNAD J. Edu*, *3*, 91–101.

Brown, P., & Lauder, H. (1996). Education, globalization and economic development. *Journal of Education Policy*, *11*(1), 1–25. doi:10.1080/0268093960110101

CAA. (2017). *African textile*. Contemporary-african-art.com.

Cabillan, M. L. (2011). *Changing landscape of mathematics in the Philippine: Lesson from globalization*. Retrieved 1/3/2017 from http://vlir-piuc.slu.edu.ph

Cabillan, M. L. (2011). *Changing landscape of mathematics in the Philippine: lesson from globalization*. Retrieved 9th May 2012 from http://vlir-piuc.slu.edu.ph

Carnoy, M. (1999). Globalization and education reform: what planners need to know. UNESCO/IIEP.

Carter, L. (2007). Sociocultural Influences on Science Education: Innovation for Contemporary Times. Wiley InterScience. *Science Education.*, 166–181. doi:10.1002ce

Carter, L. (2008). Globalization and science education: The implications of science in the new economy. *Journal of Research in Science Teaching*, *45*(5), 617–633. doi:10.1002/tea.20189

Carvus, N., & Alhih, M. S. (2014). Learning management systems use in science education. *Procedia: Social and Behavioral Sciences*, *143*, 517–520. doi:10.1016/j.sbspro.2014.07.429

Cassell, J., & Jenkins, H. (Eds.). (1998). *From Barbie to mortal kombat: Gender and computer games*. Cambridge, MA: MIT Press.

Cassie, Q. (2009). Globalization and Science Education: The Implications for Indigenous Knowledge Systems. *International Education Studies*, *2*(1), 76–88. Retrieved from www.ccsenet.org/journal.html

Charlton, B., & Andras, P. (2003). *The modernization imperative*. Exeter, UK: Imprint Academic.

Christie, K. (2005, April). Virginia's Excellent Adventure. *Phi Delta Kappan.*

Chuang, T., & Chen, W. (2009). Effect of computer-based video games on children: An experimental study. *Journal of Educational Technology & Society, 12*(2), 1–10.

Clarkson, P., & Atweh, B. (2004). *More Perspectives on the Impact of Globalisation on Mathematics Education in Higher Education in Australasia.* Australian Research Council Large Research Grant.

Confrey, J. (2004). Undertaking data analysis of student outcomes as professional development for teachers. *ZDM, 36*(1), 1–9.

Conklin, H. C. (1962) Lexicographical treatment of folk taxonomies. International Journal of American Linguistics, 28(2), 119-141.

Cooper, J. (2006). The digital divide: The special case of gender. *Journal of Computer Assisted Learning, 22*(5), 320–334. doi:10.1111/j.1365-2729.2006.00185.x

Crommey, A. (2000). *Using Student Assessment Data: What Can We Learn From Schools?* Oak Brook, IL: North Central Regional Educational Laboratory.

D'Ambrosio, U. (1985). Ethnomathematics and its place in the history and pedagogy of mathematics. *For the Learning of Mathematics, 5*(1), 44–48.

D'Ambrosio, U. (1990). *Etnomatemática* [Ethnomathematics]. São Paulo: Editora Ática.

D'Ambrosio, U. (1993). Ethnomathematics, the nature of mathematics and mathematics education. In P. Ernest (Ed.), *Mathematics, education and philosophy: An international perspective* (pp. 230–242). London: Falmer Press.

D'Ambrosio, U. (1999). Literacy, Matheracy, and Technoracy: A Trivium for Today. *Mathematical Thinking and Learning, 1*(2), 131–153. doi:10.120715327833mtl0102_3

David, S. A., & Wildemeersch, D. (2007). *Globalization and Curriculum Restructuring in Higher Education: Comparing the impacts on Higher Education Curriculum Planning in the States Kerala and Tamil Nadu of India (Draft doctoral proposal).* Katholieke Universiteit Leuven.

de Abreu, G. (2002). Mathematics learning in out-of-school contexts: A cultural psychology perspective. In L. D. English (Ed.), *Handbook of international research in mathematics education* (pp. 323–353). Mahwah, NJ: Lawrence Erlbaum Associates, Inc.

Denvir, B. (1988). What are we assessing in mathematics and what are we assessing for? In D. Pimm (Ed.), *Mathematics and teacher education* (pp. 128–140). London: Hodder & Stoughton.

Dlaska, A. (2000). Integrating culture and language learning in institution-wide language programmes. *Language, Culture and Curriculum, 13*(3), 247–263. doi:10.1080/07908310008666602

Dooly, M., & Villanueva, M. (2006). Internationalisation as a key dimension to teacher Education. *European Journal of Teacher Education, 29*(2), 223–240. doi:10.1080/02619760600617409

Dowling, P. (1998). *The sociology of mathematics education: Mathematical myths/ pedagogic texts*. London: Falmer Press.

Downes, T. (1999). Playing with computing technologies in the home. *Education and Information Technologies, 4*(1), 65–79. doi:10.1023/A:1009607432286

Ebrahim, A. (2010). *Characteristics of Modern Mathematics*. Mathematical Science & Technologies.

Ernest, P. (2007). *Globalization*. Ideology and Research in Mathematics Education.

Etuk, E. D., & Bello, D. O. (n.d.). Challenges and prospects of mathematics Education in Nigeria. *Journal of Assertiveness*. Retrieved from: http://www.globalacademicgroup.com/journals/assertiveness/Emmanuel%20Dan.pdf

Facer, K. (2003). *Computer games and learning*. Retrieved from http://www.futurelab.org.uk/resources/documents/discussion_papers/Computer_Games_and_Learning_discpaper.pdf

Federal Government of Nigeria. (2004). *National Policy on Education*. Lagos: Federal Ministry of Education.

Federal Republic of Nigeria. (2013). National Policy on Education (6th ed.). Lagos: Nigerian Educational Research and Development Council (NERDC).

Frake, C. O. (1962). The ethnographic study of cognitive systems. In Anthropology and Human Behavior. Washington, DC: Anthropological Society.

Franklin, K. J. (1996). *K.L Pike on etic vs emic: A review and interview*. Retrieved from www.sil.org/klp/karlintv.htm

French, D. (1963). The relationship of anthropology to studies in perception and cognition. In S. Koch (Ed.), *Psychology: A Study of a Science* (Vol. 6, pp. 388–428). New York: McGraw-Hill. doi:10.1037/10590-007

Gardner, H. (2003). Multiple intelligences after twenty years. American Educational Research Association, 21.

Gates, P., & Vistro-Yu, C. P. (2003). Is mathematics for all? In *Second international handbook of mathematics education, Part one* (pp. 31–73). Dordrecht: Kluwer Academic Publishers. doi:10.1007/978-94-010-0273-8_3

Gee, J. P. (2003). *What video games have to teach us about learning and literacy*. New York: Palgrave Macmillan.

Gee, J. P. (2005). What Would a State of the Art Instructional Game Look Like? *Innovate (North Miami Beach, Fla.)*, *1*(6).

Gerdes, P. (1997). On culture, geometrical thinking and mathematics education. In A. B. Powell & M. Frankenstein (Eds.), *Ethnomathematics: Challenging eurocentrism in mathematics education* (pp. 223–247). Albany, NY: State University of New York Press.

Goodenough, W. H. (1957). Cultural anthropology and linguistics. In *Report of the 7th Annual Round Table Meeting on Linguistics and Language Study*. Georgetown University.

Griffin, S. (2007a). Early intervention for children at risk of developing mathematical learning difficulties. In D. B. Berch & M. M. M. Mazzocco (Eds.), *Why is math so hard for some children? The nature and origins of mathematical learning difficulties and disabilities* (pp. 373–395). Baltimore, MD: Paul H. Brooks Publishing Co.

Guskey, T. (2005, September). A Historical Perspective on Closing the Achievement Gap. *NASSP Bulletin*.

Hamilton, L. S. (2009). *Making sense of test-based accountability in education*. Santa Monica, CA: Rand.

Haramati, A. (2000). Teaching physiology: Filling a bucket or lighting a fire? *The Physiologist*, *43*, 117–121. PMID:10916409

Harris, S. (1999). Secondary school students' use of computers at home. *British Journal of Educational Technology*, *30*(4), 331–339. doi:10.1111/1467-8535.00123

Herz, J. C. (1997). *Joystick Nation. How videogames ate our quarters, won our hearts, and rewired our minds*. Princeton, NJ: Little Brown & Company.

Hodson, D. (2011). *Looking to the Future: Building a Curriculum for Social Activism*. Rotterdam, The Netherlands: Sense Publishers. doi:10.1007/978-94-6091-472-0

Hoffmann, F. W., & William, G. B. (1994). *Fashion & merchandising fads*. New York: Haworth Press.

Hofstede, G. J., Pedersen, P., & Hofstede, G. (2002). *Exploring culture: Exercises, stories and synthetic cultures*. Nicholas Brealey Publishing.

Holland, W., Jenkins, H., & Squire, K. (2003). Theory by Design. In M. J. P. Wolf & B. Perron (Eds.), *The video game theory reader* (p. xxii). New York: Routledge.

Hollins, E. R., & King, J. E. (1994). *Teaching diverse populations: Formulating a knowledge base* (Vol. 1944). SUNY Press.

House, J. D., & Telese, J. A. (2008). Relationships between Student and Instructional Factors and Algebra Achievement of Students in the United States and Japan: An analysis of TIMSS 2003 Educat. *Research Evaluation*, *14*(1), 101–112. doi:10.1080/13803610801896679

Hwang, G. J., Wu, P. H., & Chen, C. C. (2015). An online game approach for improving students' learning performance in web-based problem-solving activities. *Computers & Education*, *59*(1), 1246–1256.

Irish, C., & Scrubb, M. (2013). *Five competencies for culturally competent teaching and learning*. Retrieved from http://www.facultyfocus.com/articles/teaching-and-learning/five-competencies-for-culturally-competent-teaching-and-learning

Jadhav, V. N. (2012). The Role of Teacher Education in the context of Globalization. *Global online Electronic International Interdisciplinary Research Journal, 1*(1).

Jamit Software. (1999 – 2003). Jamit Fraction Games.

Joseph, G. G. (2000). *The crest of the peacock. Non-European roots of mathematics*. Princeton, NJ: Princeton University Press.

Kachur, J., & Harrison, T. (1999). Public education, globalization, and democracy: Whither Alberta? In T. Harrison & J. Kachur (Eds.), *Contested classrooms* (pp. xiii–xxxiv). Edmonton, Canada: University of Alberta.

Ke, F. (2008). Computer games application within alternative classroom goal structures: Cognitive, metacognitive, and affective evaluation. *Educational Technology Research and Development*, 56.

Kosoko-Oyedeko, G. A. (2012). Effects of Instructional Games on Primary School Pupils' Learning Outcomes in Christian Religious Studies in Lagos State, Nigeria. A P.hD post-field seminar paper presented to the Department of Teacher Education, University of Ibadan, Ibadan, Nigeria.

Kukogoh, I. S. (2015). *Pulse newsletter 15:09 11:08.2015. Pulse.ng*. WAEC.

Lerman, S. (1994). Cultural perspectives on the mathematics classroom. Dordrecht: Kluwer Academic Publishers.

Lounsbury, F. G. (1960). Similarity and contiguity relations in language and in culture. In *Report on the 10th Annual Round Table Meeting on Linguistics and Language Studies*. Georgetown University.

Lucas, K., & Sherry, J. L. (2004). Sex differences in video game play: A communication-based explanation. *Communication Research, 31*(5), 499–523. doi:10.1177/0093650204267930

Luhmann, N. (1995). *Social systems*. Cambridge, MA: Harvard University Press.

Luhmann, N. (1997). Globalization or world society: How to conceive of modern society. *International Review of Sociology, 7*(1), 67–80. doi:10.1080/03906701.1997.9971223

Makonye, P. J. (2012). Learner errors on calculus tasks in the NSC examinations: Towards an analytical protocol for learner perturbable concepts in introductory differentiation. *International Journal of Learning, 18*(6), 339–357.

Mansilla, V., & Jackson, A. (2011). *Educating for global competence*. New York, NY: Asia Society.

Martin, M., & Vaughn, B. (2007). Cultural competence: The nuts and bolts of diversity and nclusion. *Strategic Diversity & Inclusion Management, 1*(1), 31–38.

McFarlane, A., Sparrowhawk, A., & Heald, Y. (2002). *Report on the educational use of games*. Cambridge: Teachers Evaluating Educational Multimedia.

McKinley, E. (2005). Locating the global: Culture, language and science education for indigenous students. *International Journal of Science Education*, *27*(2), 227–241. doi:10.1080/0950069042000325861

Meehan, P. (1980). Science, enthnoscience, and agricultural knowledge utilization. In D. M. Warren, D. Brokensha, & O. Werner (Eds.), *Indigenous knowledge systems and development* (pp. 383–391). Lanham, MD: University Press of America.

Miller, C. S., Lehman, J. F., & Koedinger, K. R. (1999). Goals and learning in microworlds. *Cognitive Science*, *23*(3), 305–336. doi:10.120715516709cog2303_2

Miller, P. (2001). Learning styles: The multimedia of the mind. *ERIC Document ED*, *451*, 140.

Mohrman, K., Mab, W., & Baker, D. (2008). The research university in transition: The emerging global model. *Higher Education Policy*, *21*(1), 5–27. doi:10.1057/palgrave.hep.8300175

Moschkovich, J.N., & Brenner, M.E. (2002). Preface: Everyday and academic mathematics in the classroom. *Journal of Research in Mathematics Education Monograph.*

Mumtaz, S. (2001). Children's enjoyment and perception of computer use in the home and the school. *Computers & Education*, *36*(4), 347–362. doi:10.1016/S0360-1315(01)00023-9

Namukasa, I. (n.d.). School Mathematics in the Era of Globalization. *University of Alberta.*

Nayak, P. (2012). *Why do a teacher use technology in teaching mathematics*. Retrieved from https://www.slideshare.net/pratimanayak39/why-do-a-tecaher-use-technology-in- teaching-mathematics Retrieved on 1/3/2017

O'Donoghue, J. (2000). Adults, mathematics, culture and society. In D. Coben (Ed.), *Perspectives on adults learning mathematics* (pp. 101–107). Dordrecht: Kluwer Academic Publishers.

Oblinger, D. (2004). The next generation of educational engagement. *Journal of Interactive Media in Education, 2004*(8), 1-18.

Oblinger, D. G., & Verville, A. L. (1998). Global Interdependence. In *What Business wants from Higher Education. American Council of Education, Series on Higher Education*. ORYX Press.

Obomanu, B. J., & Adaramola, M. O. (2011). Factor related to under achievement in science technology and mathematics education (STME) in secondary school in Rivers State, Nigeria. *World Journal of Education, 1*(1). Retrieved 14th November, 2011 from www.sciedu.ca

Odili, G. (2006). *Mathematics in Nigeria secondary schools: A teaching perspective* (Rev.ed.). Rex Charles & Patrick Limited.

Odogwu, H. N. (1994). Primary school Teachers and the Teaching of Time Concept in Schools. *Education Today*, *7*(21), 24–30.

Ohuche, R. O. (1978). Recent attempts at Mathematics Curriculum Renewal in English Speaking West Africa. *Abacus*, *12*(1), 17–26.

Okafor, C. F., & Anaduaka, U. S. (2015). *Nigerian School Children and Mathematics Phobia: How Mathematics Teachers can help*. Journal of Science and Education Publishing.

Olaniyan, D. A., & Obadara, O. E. (2006). Globalisation and Nigerian educational System: Opportunities and challenges. *International Journal of African and African American Studies*, *5*(2). Retrieved from https://ojcs.siue.edu

Oshibodu, B. M. (1984). An Investigation of Teacher's Instrumental problems in Mathematics. *Benin Journal of Educational Studies.*, *11*(3), 43–54.

Oyen, A., & Bebko, J. M. (1996). The Effects of Computer Games and Lesson Contexts on Children's Mnemonic Strategies. *Journal of Experimental Child Psychology*, *62*(2), 173–189. doi:10.1006/jecp.1996.0027 PMID:8812043

Penn, H. (2005). *Unequal Childhoods: Young children's lives in poor countries*. London: Routledge. doi:10.4324/9780203465349

Pewewardy, C., & Hammer, P. C. (2003). *Culturally Responsive Teaching for American Indian Students*. ERIC Digest.

Pike, K. (1954). *Emic and etic standpoints for the description of behaviour*. Glendale, IL: Summer Institute of Linguistics.

Pinxten, R. (1987). Applications in the teaching of mathematics and sciences. In A. B. Powell & M. Frankenstein (Eds.), *Ethnomathematics: Challenging eurocentrism in mathematics education* (pp. 373–399). Albany, NY: State University of New York Press.

Pinxten, R. (1991). Some remarks on Navajo geometry and Piagetian genetic theory. *Infancia y Aprendizaje*, *54*(54), 41–52. doi:10.1080/02103702.1991.10822292

Pope, D. C. (2001). *Doing school: How we are creating a generation of stressed out, materialistic, and miseducated students*. Yale University Press.

Popoola, A. A. (2004). *Effects of programmed instructional strategies on Senior Secondary School student learning outcomes in Mathematics in Ekiti State, Nigeria* (An unpublished Ph.D Thesis). University of Ibadan.

Rampal, A. (2003). Counting on every day mathematics. In T. S. Saraswati (Ed.), *Crosscultural perspectives in human development: theory, research, and applications* (pp. 326–353). New Delhi: Sage Publications.

Remis, K. K. (2015). *LMS enhances K12 instruction: Systems increase engagement, provide quick access to digital resources and help teachers with administrative tasks*. Academic Press.

Rice, J. (2007). New media resistance: Barriers to implementation of computer video games in the classroom. 668. *Journal of Educational Multimedia and Hypermedia*, *16*(3), 249–261.

Rosemberg, C. R., Stein, A., & Alam, F. (2013). At home and at school: Bridging literacy for children from poor rural or marginalized urban communities. International Handbook of Research on Children's Literacy, Learning, and Culture, 67-82.

Rutkwoski, L., & Rutkwoski, D. (2008). *Globalization in Education: An empirical look at globalization in education. An Example with Trends in International Mathematics and Science Study Mathematics data.* Retrieved 9th May, 2012 from www.iea.nl

Salami, I. A (2014). *Effects of Training Programmes o Pre-service Primary Mathematics Teachers' Planning and Utilization of Activity-Based Lessons in Southwestern States, Nigeria*. Unpublished Ph.D. theses.

Salman, M. F. (2005). Teachers' identification of the difficulty levels of topics in the primary school mathematics curriculum in Kwara state abacus. *The Journal of Mathematical Association of Nigeria*, *30*(1), 20–29.

Smith, D. G. (2000). The specific challenges of globalization for teaching and vice versa. *The Alberta Journal of Educational Research*, *46*(1).

Steen. (1998). *The Future of Mathematics Education*. Association for Supervision and Curriculum Development.

Stewart, V. (2012). *World-Class Education. Learning from international of world of experience and innovation*. Retrieved from www.ascd.org

Sturtevant, C. W. (2009). Anthropological Approaches. Studies in Ethnoscience. *American Anthropologist*, *66*(3). doi:10.1525/aa.1964.66.3.02a00850

Sun, D., Ma, N., Bao, M., Chen, X.-C., & Zhang, D.-R. (2008). Computer Games: A Double Edged Sword? *Cyberpsychology & Behavior*, *11*(5), 545–548. doi:10.1089/cpb.2007.0145 PMID:18817481

Tella, A. (1998). *An Investigation into Poor Performance and Attitude of Secondary School Students towards Mathematics in Oyo State, Nigeria*. Ibadan: Project University of Ibadan.

Texin, T. (2002). *Survey: Who Do You Save?* Retrieved: November 9, 2005. http://www.i18nguy.com/adventures/whotosave.html

Theonundku, S., Nwamuo, C., & Okinamiri, A. (2013). ICT Pedagogy Integration in teacher training. *Nigeria Journal of Professional Teachers*, *3*(1).

Udofa A. I. & Udo I. E. (n.d.). Globalisation and its influence on mathematics education in secondary schools in Akwa Ibom state, Nigeria. *Journal of Educational and Social Research, 3*(6), 33-40.

Udofa, A. I., & Udo, I. E. (2013). Globalisation and Its Influence on Mathematics Education in Secondary Schools in Akwa Ibom State, Nigeria. *Journal of Educational and Social Research*, *3*(6).

Ugboduma S. O. (2008). Introducing Mathematics laboratory into secondary schools Mathematics curriculum to promote students' enthusian for natural development. *Journal of Mathematical Association of Nigeria,* 187-196.

Ugwu, P. N. (2013). Mathematics as a tool for rebranding Nigeria: Implication for secondary Mathematic education. *Abacus*, *38*(11).

Ukeje, B. O. (1990). *Teaching Mathematics in Secondary School*. Anambra Anachuna Educational Books.

Umeasiegbu, O., & Esomonu, N. (2012). E-Learning in Global Education: Challenges and Prospects for Science and Physical Education In Nigeria. *Journal of Education and Practice*, *3*(14), 44–48.

University of Washington. (2011). Gender stereotypes about math develop as early as second grade. *ScienceDaily*.

Varghese, T., & McCusker, D. P. (2006). On Globalization and Ethnomathematics. *Canadian and International Education /Education canadienne et international*, *35*(1), 2. Available at: http://ir.lib.uwo.ca/cie-eci/vol35/iss1/2

Velom, V., & Rohani, F. L. (2010). Interactive Science Homework. An Experiment in Home and School Connections. *National Association of Secondary School Principal Bulleting*, *85*(2), 20–32.

Vernille, K. (2002). *Why are mathematics students falling behind their international peers*. National Education Policy Center. Retrieved 19th June, 2015 from www.users.math.umd.edu

Vithal, R., & Skovsmose, O. (1997). The end of innocence: A critique of ethnomathematics. *Educational Studies in Mathematics*, *34*(2), 131–157. doi:10.1023/A:1002971922833

Vogel, J. J., Vogel, D. S., Cannon-Bowers, J., Bowers, C. A., Muse, K., & Wright, M. (2006). Computer gaming and interactive simulations for learning: A meta-analysis. *Journal of Educational Computing Research*, *34*(3), 229–243. doi:10.2190/FLHV-K4WA-WPVQ-H0YM

Vulliamy, G. (2004). The impact of Globalization on qualitative research in comparative and international education. *Compare: A Journal of Comparative Education*, *34*(3), 261–284. doi:10.1080/0305792042000257112

Vulliamy, G., & Webb, R. (2006). Education policy under New Labour. In R. Webb (Ed.), *Changing teaching and learning in the primary school*. Buckingham, UK: Open University Press.

Wildavsky, B. (2010). *The great brain race: How global universities are reshaping the world.* Princeton, NJ: Princeton University Press.

Wokocha, A. M. (2011). *Globalisation, education reforms and knowledge economy in a multicultural society*. First National Conference, Institute of Education, University of Port Harcourt. Retrieved from www.goeicrij.com

Wokocha, A. M. (2011). Globalisation, education reforms and knowledge economy in a multicultural society. *First National Conference*, Institute of Education, University of Port Harcourt.

Würtz, E. (2005). A cross-cultural analysis of websites from high-context cultures and low-context cultures. *Journal of Computer-Mediated Communication*, *11*(1), 13. doi:10.1111/j.1083-6101.2006.tb00313.x

Yang, T. C. (2012). Building virtual cities, inspiring intelligent citizens: Digital games for developing students' problem solving and learning motivation. *Computers & Education*, *59*(2), 365–377. doi:10.1016/j.compedu.2012.01.012

Yara, N. (2009). *Mathematics skills and students achievement.* Oregon Annual Reports.

Yeh, S. (2006, April). High-Stakes Testing: "Can Rapid Assessment Reduce the Pressure?" *Teacher College Record.*

Related References

To continue our tradition of advancing research in the field of education, we have compiled a list of recommended IGI Global readings. These references will provide additional information and guidance to further enrich your knowledge and assist you with your own research and future publications.

Abrami, P. C., Savage, R. S., Deleveaux, G., Wade, A., Meyer, E., & LeBel, C. (2010). The learning toolkit: The design, development, testing and dissemination of evidence-based educational software. In P. Zemliansky & D. Wilcox (Eds.), *Design and implementation of educational games: Theoretical and practical perspectives* (pp. 168–188). Hershey, PA: IGI Global. doi:10.4018/978-1-61520-781-7.ch012

Ackfeldt, A., & Malhotra, N. (2012). Do managerial strategies influence service behaviours? Insights from a qualitative study. In R. Eid (Ed.), *Successful customer relationship management programs and technologies: Issues and trends* (pp. 174–187). Hershey, PA: IGI Global. doi:10.4018/978-1-4666-0288-5.ch013

Adi, A., & Scotte, C. G. (2013). Barriers to emerging technology and social media integration in higher education: Three case studies. In M. Pătruţ & B. Pătruţ (Eds.), *Social media in higher education: Teaching in web 2.0* (pp. 334–354). Hershey, PA: IGI Global. doi:10.4018/978-1-4666-2970-7.ch017

Aldana-Vargas, M. F., Gras-Martí, A., Montoya, J., & Osorio, L. A. (2013). Pedagogical counseling program development through an adapted community of inquiry framework. In Z. Akyol & D. Garrison (Eds.), *Educational communities of inquiry: Theoretical framework, research and practice* (pp. 350–373). Hershey, PA: IGI Global. doi:10.4018/978-1-4666-2110-7.ch017

Alegre, O. M., & Villar, L. M. (2011). Faculty professional learning: An examination of online development and assessment environments. In G. Vincenti & J. Braman (Eds.), *Teaching through multi-user virtual environments: Applying dynamic elements to the modern classroom* (pp. 66–93). Hershey, PA: IGI Global. doi:10.4018/978-1-61692-822-3.ch006

Alegre, O. M., & Villar, L. M. (2012). Faculty professional learning: An examination of online development and assessment environments. In *Organizational learning and knowledge: Concepts, methodologies, tools and applications* (pp. 305–331). Hershey, PA: IGI Global. doi:10.4018/978-1-60960-783-8.ch119

Alegre-Rosa, O. M., & Villar-Angulo, L. M. (2010). Training of teachers in virtual scenario: An excellence model for quality assurance in formative programmes. In S. Mukerji & P. Tripathi (Eds.), *Cases on transnational learning and technologically enabled environments* (pp. 190–213). Hershey, PA: IGI Global. doi:10.4018/978-1-61520-749-7.ch011

Andegherghis, S. (2012). Technology and traditional teaching. In I. Chen & D. McPheeters (Eds.), *Cases on educational technology integration in urban schools* (pp. 74–79). Hershey, PA: IGI Global. doi:10.4018/978-1-61350-492-5.ch019

Anderson, K. H., & Muirhead, W. (2013). Blending storytelling with technology in the professional development of police officers. In H. Yang & S. Wang (Eds.), *Cases on formal and informal e-learning environments: Opportunities and practices* (pp. 143–165). Hershey, PA: IGI Global. doi:10.4018/978-1-4666-1930-2.ch008

Anderson, S., & Oyarzun, B. (2013). Multi-modal professional development for faculty. In J. Keengwe & L. Kyei-Blankson (Eds.), *Virtual mentoring for teachers: Online professional development practices* (pp. 43–65). Hershey, PA: IGI Global. doi:10.4018/978-1-4666-1963-0.ch003

Annetta, L. A., Holmes, S., & Cheng, M. (2012). Measuring student perceptions: Designing an evidenced centered activity model for a serious educational game development software. In R. Ferdig & S. de Freitas (Eds.), *Interdisciplinary advancements in gaming, simulations and virtual environments: Emerging trends* (pp. 165–182). Hershey, PA: IGI Global. doi:10.4018/978-1-4666-0029-4.ch011

Austin, R., & Anderson, J. (2006). Re-schooling and information communication technology: A case study of Ireland. In L. Tan Wee Hin & R. Subramaniam (Eds.), *Handbook of research on literacy in technology at the K-12 level* (pp. 176–194). Hershey, PA: IGI Global. doi:10.4018/978-1-59140-494-1.ch010

Ayling, D., Owen, H., & Flagg, E. (2014). From basic participation to transformation: Immersive virtual professional development. In S. Leone (Ed.), *Synergic integration of formal and informal e-learning environments for adult lifelong learners* (pp. 47–74). Hershey, PA: IGI Global. doi:10.4018/978-1-4666-4655-1.ch003

Ayoola, K. A. (2010). An appraisal of a computer-based continuing professional development (CPD) Course for Nigerian English teachers and teacher-trainers. In R. Taiwo (Ed.), *Handbook of research on discourse behavior and digital communication: Language structures and social interaction* (pp. 642–650). Hershey, PA: IGI Global. doi:10.4018/978-1-61520-773-2.ch041

Baia, P. (2011). The trend of commitment: Pedagogical quality and adoption. In S. D'Agustino (Ed.), *Adaptation, resistance and access to instructional technologies: Assessing future trends in education* (pp. 273–315). Hershey, PA: IGI Global. doi:10.4018/978-1-61692-854-4.ch017

Banas, J. R., & Velez-Solic, A. (2013). Designing effective online instructor training and professional development. In J. Keengwe & L. Kyei-Blankson (Eds.), *Virtual mentoring for teachers: Online professional development practices* (pp. 1–25). Hershey, PA: IGI Global. doi:10.4018/978-1-4666-1963-0.ch001

Banks, W. P., & Van Sickle, T. (2011). Digital partnerships for professional development: Rethinking university–public school collaborations. In M. Bowdon & R. Carpenter (Eds.), *Higher education, emerging technologies, and community partnerships: Concepts, models and practices* (pp. 153–163). Hershey, PA: IGI Global. doi:10.4018/978-1-60960-623-7.ch014

Barbour, M. K., Siko, J., Gross, E., & Waddell, K. (2013). Virtually unprepared: Examining the preparation of K-12 online teachers. In R. Hartshorne, T. Heafner, & T. Petty (Eds.), *Teacher education programs and online learning tools: Innovations in teacher preparation* (pp. 60–81). Hershey, PA: IGI Global. doi:10.4018/978-1-4666-1906-7.ch004

Bartlett, J. E. II, & Bartlett, M. E. (2009). Innovative strategies for preparing and developing career and technical education leaders. In V. Wang (Ed.), *Handbook of research on e-learning applications for career and technical education: Technologies for vocational training* (pp. 248–261). Hershey, PA: IGI Global. doi:10.4018/978-1-60566-739-3.ch020

Baylen, D. M., & Glacken, J. (2007). Promoting lifelong learning online: A case study of a professional development experience. In Y. Inoue (Ed.), *Online education for lifelong learning* (pp. 229–252). Hershey, PA: IGI Global. doi:10.4018/978-1-59904-319-7.ch011

Beedle, J., & Wang, S. (2013). Roles of a technology leader. In S. Wang & T. Hartsell (Eds.), *Technology integration and foundations for effective leadership* (pp. 228–241). Hershey, PA: IGI Global. doi:10.4018/978-1-4666-2656-0.ch013

Begg, M., Dewhurst, D., & Ross, M. (2010). Game informed virtual patients: Catalysts for online learning communities and professional development of medical teachers. In J. Lindberg & A. Olofsson (Eds.), *Online learning communities and teacher professional development: Methods for improved education delivery* (pp. 190–208). Hershey, PA: IGI Global. doi:10.4018/978-1-60566-780-5.ch011

Begg, M., Dewhurst, D., & Ross, M. (2010). Game informed virtual patients: Catalysts for online learning communities and professional development of medical teachers. In R. Luppicini & A. Haghi (Eds.), *Cases on digital technologies in higher education: Issues and challenges* (pp. 304–322). Hershey, PA: IGI Global. doi:10.4018/978-1-61520-869-2.ch020

Benton, C. J., White, O. L., & Stratton, S. K. (2014). Collaboration not competition: International education expanding perspectives on learning and workforce articulation. In V. Wang (Ed.), *International education and the next-generation workforce: Competition in the global economy* (pp. 64–82). Hershey, PA: IGI Global. doi:10.4018/978-1-4666-4498-4.ch004

Benus, M. J., Yarker, M. B., Hand, B. M., & Norton-Meier, L. A. (2013). Analysis of discourse practices in elementary science classrooms using argument-based inquiry during whole-class dialogue. In M. Khine & I. Saleh (Eds.), *Approaches and strategies in next generation science learning* (pp. 224–245). Hershey, PA: IGI Global. doi:10.4018/978-1-4666-2809-0.ch012

Betts, K., Kramer, R., & Gaines, L. L. (2013). Online faculty and adjuncts: Strategies for meeting current and future demands of online education through online human touch training and support. In M. Raisinghani (Ed.), *Curriculum, learning, and teaching advancements in online education* (pp. 94–112). Hershey, PA: IGI Global. doi:10.4018/978-1-4666-2949-3.ch007

Biesinger, K. D., & Crippen, K. J. (2010). Designing and delivering technology integration to engage students. In J. Yamamoto, J. Leight, S. Winterton, & C. Penny (Eds.), *Technology leadership in teacher education: Integrated solutions and experiences* (pp. 298–313). Hershey, PA: IGI Global. doi:10.4018/978-1-61520-899-9.ch016

Bledsoe, C., & Pilgrim, J. (2013). Three instructional models to integrate technology and build 21st century literacy skills. In J. Whittingham, S. Huffman, W. Rickman, & C. Wiedmaier (Eds.), *Technological tools for the literacy classroom* (pp. 243–262). Hershey, PA: IGI Global. doi:10.4018/978-1-4666-3974-4.ch014

Bloom, L., & Dole, S. (2014). Virtual school of the smokies. In S. Mukerji & P. Tripathi (Eds.), *Handbook of research on transnational higher education* (pp. 674–689). Hershey, PA: IGI Global. doi:10.4018/978-1-4666-4458-8.ch034

Bober, M. J. (2005). Ensuring quality in technology-focused professional development. In C. Howard, J. Boettcher, L. Justice, K. Schenk, P. Rogers, & G. Berg (Eds.), *Encyclopedia of distance learning* (pp. 845–852). Hershey, PA: IGI Global. doi:10.4018/978-1-59140-555-9.ch121

Bober, M. J. (2009). Ensuring quality in technology-focused professional development. In P. Rogers, G. Berg, J. Boettcher, C. Howard, L. Justice, & K. Schenk (Eds.), *Encyclopedia of distance learning* (2nd ed.; pp. 924–931). Hershey, PA: IGI Global. doi:10.4018/978-1-60566-198-8.ch129

Boling, E. C., & Beatty, J. (2012). Overcoming the tensions and challenges of technology integration: How can we best support our teachers? In R. Ronau, C. Rakes, & M. Niess (Eds.), *Educational technology, teacher knowledge, and classroom impact: A research handbook on frameworks and approaches* (pp. 136–156). Hershey, PA: IGI Global. doi:10.4018/978-1-60960-750-0.ch006

Bovard, B., Bussmann, S., Parra, J., & Gonzales, C. (2010). Transitioning to e-learning: Teaching the teachers. In *Web-based education: Concepts, methodologies, tools and applications* (pp. 259–276). Hershey, PA: IGI Global. doi:10.4018/978-1-61520-963-7.ch019

Bowskill, N. (2009). Informal learning projects and world wide voluntary co-mentoring. In P. Rogers, G. Berg, J. Boettcher, C. Howard, L. Justice, & K. Schenk (Eds.), *Encyclopedia of distance learning* (2nd ed.; pp. 1169–1177). Hershey, PA: IGI Global. doi:10.4018/978-1-60566-198-8.ch167

Bowskill, N., & McConnell, D. (2010). Collaborative reflection in globally distributed inter-cultural course teams. In G. Berg (Ed.), *Cases on online tutoring, mentoring, and educational services: Practices and applications* (pp. 172–184). Hershey, PA: IGI Global. doi:10.4018/978-1-60566-876-5.ch014

Bradley, J. B., Rachal, J., & Harper, L. (2013). Online professional development for adults: Utilizing andragogical methods in research and practice. In V. Bryan & V. Wang (Eds.), *Technology use and research approaches for community education and professional development* (pp. 171–193). Hershey, PA: IGI Global. doi:10.4018/978-1-4666-2955-4.ch011

Braun, P. (2013). Clever health: A study on the adoption and impact of an ehealth initiative in rural Australia. In M. Cruz-Cunha, I. Miranda, & P. Gonçalves (Eds.), *Handbook of research on ICTs and management systems for improving efficiency in healthcare and social care* (pp. 69–87). Hershey, PA: IGI Global. doi:10.4018/978-1-4666-3990-4.ch004

Breen, P. (2014). An intramuscular approach to teacher development in international collaborative higher education. In S. Mukerji & P. Tripathi (Eds.), *Handbook of research on transnational higher education* (pp. 368–390). Hershey, PA: IGI Global. doi:10.4018/978-1-4666-4458-8.ch019

Brown, C. A., & Neal, R. E. (2013). Definition and history of online professional development. In J. Keengwe & L. Kyei-Blankson (Eds.), *Virtual mentoring for teachers: Online professional development practices* (pp. 182–203). Hershey, PA: IGI Global. doi:10.4018/978-1-4666-1963-0.ch010

Brown, C. W., & Peters, K. A. (2013). STEM academic enrichment and professional development programs for K-12 urban students and teachers. In R. Lansiquot (Ed.), *Cases on interdisciplinary research trends in science, technology, engineering, and mathematics: Studies on urban classrooms* (pp. 19–56). Hershey, PA: IGI Global. doi:10.4018/978-1-4666-2214-2.ch002

Burgess, M. (2013). Using second life to support student teachers' socio-reflective practice: A mixed-method analysis. In R. Lansiquot (Ed.), *Cases on interdisciplinary research trends in science, technology, engineering, and mathematics: Studies on urban classrooms* (pp. 107–127). Hershey, PA: IGI Global. doi:10.4018/978-1-4666-2214-2.ch006

Burner, K. J. (2012). Web 2.0, the individual, and the organization: Privacy, confidentiality, and compliance. In V. Dennen & J. Myers (Eds.), *Virtual professional development and informal learning via social networks* (pp. 25–38). Hershey, PA: IGI Global. doi:10.4018/978-1-4666-1815-2.ch002

Buzzetto-More, N. (2010). Applications of second life. In H. Song & T. Kidd (Eds.), *Handbook of research on human performance and instructional technology* (pp. 149–162). Hershey, PA: IGI Global. doi:10.4018/978-1-60566-782-9.ch009

Bynog, M. (2013). Development of a technology plan. In S. Wang & T. Hartsell (Eds.), *Technology integration and foundations for effective leadership* (pp. 88–101). Hershey, PA: IGI Global. doi:10.4018/978-1-4666-2656-0.ch006

Calway, B. A., & Murphy, G. A. (2011). A work-integrated learning philosophy and the educational imperatives. In P. Keleher, A. Patil, & R. Harreveld (Eds.), *Work-integrated learning in engineering, built environment and technology: Diversity of practice in practice* (pp. 1–24). Hershey, PA: IGI Global. doi:10.4018/978-1-60960-547-6.ch001

Carlén, U., & Lindström, B. (2012). Informed design of educational activities in online learning communities. In A. Olofsson & J. Lindberg (Eds.), *Informed design of educational technologies in higher education: Enhanced learning and teaching* (pp. 118–134). Hershey, PA: IGI Global. doi:10.4018/978-1-61350-080-4.ch007

Cassidy, A., Sipos, Y., & Nyrose, S. (2014). Supporting sustainability education and leadership: Strategies for students, faculty, and the planet. In S. Mukerji & P. Tripathi (Eds.), *Handbook of research on transnational higher education* (pp. 207–231). Hershey, PA: IGI Global. doi:10.4018/978-1-4666-4458-8.ch012

Cavanagh, T. B. (2011). Leveraging online university education to improve K-12 science education: The ScienceMaster case study. In M. Bowdon & R. Carpenter (Eds.), *Higher education, emerging technologies, and community partnerships: Concepts, models and practices* (pp. 221–233). Hershey, PA: IGI Global. doi:10.4018/978-1-60960-623-7.ch020

Chapman, D. L. (2013). Overview of technology plans. In S. Wang & T. Hartsell (Eds.), *Technology integration and foundations for effective leadership* (pp. 71–87). Hershey, PA: IGI Global. doi:10.4018/978-1-4666-2656-0.ch005

Chapman, D. L., Bynog, M., & Yocom, H. (2013). Assessment, evaluation, and revision of a technology plan. In S. Wang & T. Hartsell (Eds.), *Technology integration and foundations for effective leadership* (pp. 124–150). Hershey, PA: IGI Global. doi:10.4018/978-1-4666-2656-0.ch008

Chylinski, R., & Hanewald, R. (2009). Creating supportive environments for CALL teacher autonomy. In R. de Cássia Veiga Marriott & P. Lupion Torres (Eds.), *Handbook of research on e-learning methodologies for language acquisition* (pp. 387–408). Hershey, PA: IGI Global. doi:10.4018/978-1-59904-994-6.ch024

Chylinski, R., & Hanewald, R. (2011). Creating supportive environments for CALL teacher autonomy. In *Instructional design: Concepts, methodologies, tools and applications* (pp. 840–860). Hershey, PA: IGI Global. doi:10.4018/978-1-60960-503-2.ch403

Clouse, N. K., Williams, S. R., & Evans, R. D. (2011). Developing an online mentoring program for beginning teachers. In S. D'Agustino (Ed.), *Adaptation, resistance and access to instructional technologies: Assessing future trends in education* (pp. 410–428). Hershey, PA: IGI Global. doi:10.4018/978-1-61692-854-4.ch024

Colomo-Palacios, R., Tovar-Caro, E., García-Crespo, Á., & Gómez-Berbís, J. M. (2012). Identifying technical competences of IT professionals: The case of software engineers. In R. Colomo-Palacios (Ed.), *Professional advancements and management trends in the IT sector* (pp. 1–14). Hershey, PA: IGI Global. doi:10.4018/978-1-4666-0924-2.ch001

Corbeil, M. E., & Corbeil, J. R. (2012). Creating ongoing online support communities through social networks to promote professional learning. In V. Dennen & J. Myers (Eds.), *Virtual professional development and informal learning via social networks* (pp. 114–133). Hershey, PA: IGI Global. doi:10.4018/978-1-4666-1815-2.ch007

Corbitt, B., Holt, D., & Segrave, S. (2008). Strategic design for web-based teaching and learning: Making corporate technology systems work for the learning organization. In L. Tomei (Ed.), *Online and distance learning: Concepts, methodologies, tools, and applications* (pp. 897–904). Hershey, PA: IGI Global. doi:10.4018/978-1-59904-935-9.ch075

Corbitt, B., Holt, D. M., & Segrave, S. (2008). Strategic design for web-based teaching and learning: Making corporate technology system work for the learning organization. In L. Esnault (Ed.), *Web-based education and pedagogical technologies: Solutions for learning applications* (pp. 280–302). Hershey, PA: IGI Global. doi:10.4018/978-1-59904-525-2.ch016

Coutinho, C. P. (2010). Challenges for teacher education in the learning society: Case studies of promising practice. In H. Yang & S. Yuen (Eds.), *Handbook of research on practices and outcomes in e-learning: Issues and trends* (pp. 385–401). Hershey, PA: IGI Global. doi:10.4018/978-1-60566-788-1.ch023

Cowie, B., Jones, A., & Harlow, A. (2011). Technological infrastructure and implementation environments: The case of laptops for New Zealand teachers. In S. D'Agustino (Ed.), *Adaptation, resistance and access to instructional technologies: Assessing future trends in education* (pp. 40–52). Hershey, PA: IGI Global. doi:10.4018/978-1-61692-854-4.ch003

Crichton, S. (2007). A great wall of difference: Musings on instructional design in contemporary China. In M. Keppell (Ed.), *Instructional design: Case studies in communities of practice* (pp. 91–105). Hershey, PA: IGI Global. doi:10.4018/978-1-59904-322-7.ch005

Croasdaile, S. (2009). Inter-organizational e-collaboration in education. In N. Kock (Ed.), *E-collaboration: Concepts, methodologies, tools, and applications* (pp. 1157–1170). Hershey, PA: IGI Global. doi:10.4018/978-1-60566-652-5.ch086

Croasdaile, S. (2009). Inter-organizational e-collaboration in education. In J. Salmons & L. Wilson (Eds.), *Handbook of research on electronic collaboration and organizational synergy* (pp. 16–29). Hershey, PA: IGI Global. doi:10.4018/978-1-60566-106-3.ch002

Csoma, K. (2010). EPICT: Transnational teacher development through blended learning. In S. Mukerji & P. Tripathi (Eds.), *Cases on technological adaptability and transnational learning: Issues and challenges* (pp. 147–161). Hershey, PA: IGI Global. doi:10.4018/978-1-61520-779-4.ch008

Cunningham, C. A., & Harrison, K. (2011). The affordances of second life for education. In G. Vincenti & J. Braman (Eds.), *Teaching through multi-user virtual environments: Applying dynamic elements to the modern classroom* (pp. 94–119). Hershey, PA: IGI Global. doi:10.4018/978-1-61692-822-3.ch007

Curwood, J. S. (2014). From collaboration to transformation: Practitioner research for school librarians and classroom teachers. In K. Kennedy & L. Green (Eds.), *Collaborative models for librarian and teacher partnerships* (pp. 1–11). Hershey, PA: IGI Global. doi:10.4018/978-1-4666-4361-1.ch001

Cuthell, J. P. (2010). Thinking things through: Collaborative online professional development. In J. Lindberg & A. Olofsson (Eds.), *Online learning communities and teacher professional development: Methods for improved education delivery* (pp. 154–167). Hershey, PA: IGI Global. doi:10.4018/978-1-60566-780-5.ch009

D'Agustino, S., & King, K. P. (2011). Access and advancement: Teacher transformation and student empowerment through technology mentoring. In S. D'Agustino (Ed.), *Adaptation, resistance and access to instructional technologies: Assessing future trends in education* (pp. 362–380). Hershey, PA: IGI Global. doi:10.4018/978-1-61692-854-4.ch021

Dana, N. F., Krell, D., & Wolkenhauer, R. (2013). Taking action research in teacher education online: Exploring the possibilities. In R. Hartshorne, T. Heafner, & T. Petty (Eds.), *Teacher education programs and online learning tools: Innovations in teacher preparation* (pp. 357–374). Hershey, PA: IGI Global. doi:10.4018/978-1-4666-1906-7.ch019

Dawson, K., Cavanaugh, C., & Ritzhaupt, A. D. (2013). ARTI: An online tool to support teacher action research for technology integration. In R. Hartshorne, T. Heafner, & T. Petty (Eds.), *Teacher education programs and online learning tools: Innovations in teacher preparation* (pp. 375–391). Hershey, PA: IGI Global. doi:10.4018/978-1-4666-1906-7.ch020

De Simone, C., Marquis, T., & Groen, J. (2013). Optimizing conditions for learning and teaching in K-20 education. In V. Wang (Ed.), *Handbook of research on teaching and learning in K-20 education* (pp. 535–552). Hershey, PA: IGI Global. doi:10.4018/978-1-4666-4249-2.ch031

Dennen, V. P., & Jiang, W. (2012). Twitter-based knowledge sharing in professional networks: The organization perspective. In V. Dennen & J. Myers (Eds.), *Virtual professional development and informal learning via social networks* (pp. 241–255). Hershey, PA: IGI Global. doi:10.4018/978-1-4666-1815-2.ch014

Dexter, S. (2002). eTIPS - Educational technology integration and implementation principles. In P. Rogers (Ed.), Designing instruction for technology-enhanced learning (pp. 56-70). Hershey, PA: IGI Global. doi:10.4018/978-1-930708-28-0.ch003

Dickerson, J., Winslow, J., & Lee, C. Y. (2013). Teacher training and technology: Current uses and future trends. In V. Wang (Ed.), *Handbook of research on technologies for improving the 21st century workforce: Tools for lifelong learning* (pp. 243–256). Hershey, PA: IGI Global. doi:10.4018/978-1-4666-2181-7.ch016

DiMarco, J. (2006). Cases and interviews. In J. DiMarco (Ed.), *Web portfolio design and applications* (pp. 222–276). Hershey, PA: IGI Global. doi:10.4018/978-1-59140-854-3.ch012

Doherty, I. (2013). Achieving excellence in teaching: A case study in embedding professional development for teaching within a research-intensive university. In D. Salter (Ed.), *Cases on quality teaching practices in higher education* (pp. 280–290). Hershey, PA: IGI Global. doi:10.4018/978-1-4666-3661-3.ch017

Donnelly, R. (2009). Transformative potential of constructivist blended problem-based learning in higher education. In C. Payne (Ed.), *Information technology and constructivism in higher education: Progressive learning frameworks* (pp. 182–202). Hershey, PA: IGI Global. doi:10.4018/978-1-60566-654-9.ch012

Donnelly, R. (2010). The nature of complex blends: Transformative problem-based learning and technology in Irish higher education. In Y. Inoue (Ed.), *Cases on online and blended learning technologies in higher education: Concepts and practices* (pp. 1–22). Hershey, PA: IGI Global. doi:10.4018/978-1-60566-880-2.ch001

Donnelly, R., & O'Farrell, C. (2006). Constructivist e-learning for staff engaged in coninuous professional development. In J. O'Donoghue (Ed.), *Technology supported learning and teaching: A staff perspective* (pp. 160–175). Hershey, PA: IGI Global. doi:10.4018/978-1-59140-962-5.ch010

Downing, K. F., & Holtz, J. K. (2008). Virtual school science. In K. Downing & J. Holtz (Eds.), *Online science learning: Best practices and technologies* (pp. 30–48). Hershey, PA: IGI Global. doi:10.4018/978-1-59904-986-1.ch003

Ehmann Powers, C., & Hewett, B. L. (2008). Building online training programs for virtual workplaces. In P. Zemliansky & K. St.Amant (Eds.), *Handbook of research on virtual workplaces and the new nature of business practices* (pp. 257–271). Hershey, PA: IGI Global. doi:10.4018/978-1-59904-893-2.ch019

Ellis, J. B., West, T. D., Grimaldi, A., & Root, G. (2013). Ernst & Young leadership and professional development center: Accounting designed for leaders. In R. Carpenter (Ed.), *Cases on higher education spaces: Innovation, collaboration, and technology* (pp. 330–355). Hershey, PA: IGI Global. doi:10.4018/978-1-4666-2673-7.ch017

Falco, J. (2008). Leading the art of the conference: Revolutionizing schooling through interactive videoconferencing. In D. Newman, J. Falco, S. Silverman, & P. Barbanell (Eds.), *Videoconferencing technology in K-12 instruction: Best practices and trends* (pp. 133–143). Hershey, PA: IGI Global. doi:10.4018/978-1-59904-331-9.ch010

Farmer, L. (2009). Fostering online communities of practice in career and technical education. In V. Wang (Ed.), *Handbook of research on e-learning applications for career and technical education: Technologies for vocational training* (pp. 192–203). Hershey, PA: IGI Global. doi:10.4018/978-1-60566-739-3.ch015

Farmer, L. (2010). Lights, camera, action! Via teacher librarian video conferencing. In S. Rummler & K. Ng (Eds.), *Collaborative technologies and applications for interactive information design: Emerging trends in user experiences* (pp. 179–188). Hershey, PA: IGI Global. doi:10.4018/978-1-60566-727-0.ch013

Farmer, L. S. (2012). Curriculum development for online learners. In V. Wang, L. Farmer, J. Parker, & P. Golubski (Eds.), *Pedagogical and andragogical teaching and learning with information communication technologies* (pp. 88–104). Hershey, PA: IGI Global. doi:10.4018/978-1-60960-791-3.ch007

Farmer, L. S. (2013). Assessment processes for online professional development. In J. Keengwe & L. Kyei-Blankson (Eds.), *Virtual mentoring for teachers: Online professional development practices* (pp. 161–180). Hershey, PA: IGI Global. doi:10.4018/978-1-4666-1963-0.ch009

Farmer, L. S. (2014). The roles of professional organizations in school library education. In V. Wang (Ed.), *International education and the next-generation workforce: Competition in the global economy* (pp. 170–193). Hershey, PA: IGI Global. doi:10.4018/978-1-4666-4498-4.ch010

Fok, A. W., & Ip, H. H. (2009). An agent-based framework for personalized learning in continuous professional development. In M. Syed (Ed.), *Strategic applications of distance learning technologies* (pp. 96–110). Hershey, PA: IGI Global. doi:10.4018/978-1-59904-480-4.ch007

Fragaki, M., & Lionarakis, A. (2011). Education for liberation: A transformative polymorphic model for ICT integration in education. In G. Kurubacak & T. Yuzer (Eds.), *Handbook of research on transformative online education and liberation: Models for social equality* (pp. 198–231). Hershey, PA: IGI Global. doi:10.4018/978-1-60960-046-4.ch011

Fuller, J. S., & Bachenheimer, B. A. (2013). Using an observation cycle for helping teachers integrate technology. In A. Ritzhaupt & S. Kumar (Eds.), *Cases on educational technology implementation for facilitating learning* (pp. 69–84). Hershey, PA: IGI Global. doi:10.4018/978-1-4666-3676-7.ch004

Gairin-Sallán, J., & Rodriguez-Gómez, D. (2010). Teacher professional development through knowledge management in educational organisations. In J. Lindberg & A. Olofsson (Eds.), *Online learning communities and teacher professional development: Methods for improved education delivery* (pp. 134–153). Hershey, PA: IGI Global. doi:10.4018/978-1-60566-780-5.ch008

Gairín-Sallán, J., & Rodríguez-Gómez, D. (2012). Teacher professional development through knowledge management in educational organisations. In *Organizational learning and knowledge: Concepts, methodologies, tools and applications* (pp. 1297–1315). Hershey, PA: IGI Global. doi:10.4018/978-1-60960-783-8.ch404

García, K., & Suzuki, R. (2008). The blended learning classroom: An online teacher training program. In M. Lytras, D. Gasevic, P. Ordóñez de Pablos, & W. Huang (Eds.), *Technology enhanced learning: Best practices* (pp. 57–80). Hershey, PA: IGI Global. doi:10.4018/978-1-59904-600-6.ch003

Gerbic, P., & Stacey, E. (2009). Conclusion. In E. Stacey & P. Gerbic (Eds.), *Effective blended learning practices: Evidence-based perspectives in ICT-facilitated education* (pp. 298–311). Hershey, PA: IGI Global. doi:10.4018/978-1-60566-296-1.ch016

Gibbons, A. N. (2010). Reflections concerning technology: A case for the philosophy of technology in early childhood teacher education and professional development programs. In S. Blake & S. Izumi-Taylor (Eds.), *Technology for early childhood education and socialization: Developmental applications and methodologies* (pp. 1–19). Hershey, PA: IGI Global. doi:10.4018/978-1-60566-784-3.ch001

Gonzales, C., Bussmann, S., Bovard, B., & Parra, J. (2007). Transitioning to e-learning: Teaching the teachers. In R. Sharma & S. Mishra (Eds.), *Cases on global e-learning practices: Successes and pitfalls* (pp. 52–72). Hershey, PA: IGI Global. doi:10.4018/978-1-59904-340-1.ch005

Gormley, P., Bruen, C., & Concannon, F. (2010). Sustainability through staff engagement: Applying a community of practice model to web 2.0 academic development programmes. In R. Donnelly, J. Harvey, & K. O'Rourke (Eds.), *Critical design and effective tools for e-learning in higher education: Theory into practice* (pp. 326–345). Hershey, PA: IGI Global. doi:10.4018/978-1-61520-879-1.ch020

Grandgenett, N., Ostler, E., Topp, N., & Goeman, R. (2012). Robotics and problem-based learning in STEM formal educational environments. In B. Barker, G. Nugent, N. Grandgenett, & V. Adamchuk (Eds.), *Robots in K-12 education: A new technology for learning* (pp. 94–119). Hershey, PA: IGI Global. doi:10.4018/978-1-4666-0182-6.ch005

Grant, M. R. (2010). Train the trainer: A competency-based model for teaching in virtual environments. In W. Ritke-Jones (Ed.), *Virtual environments for corporate education: Employee learning and solutions* (pp. 124–146). Hershey, PA: IGI Global. doi:10.4018/978-1-61520-619-3.ch008

Griffin, P., Care, E., Robertson, P., Crigan, J., Awwal, N., & Pavlovic, M. (2013). Assessment and learning partnerships in an online environment. In E. McKay (Ed.), *ePedagogy in online learning: New developments in web mediated human computer interaction* (pp. 39–54). Hershey, PA: IGI Global. doi:10.4018/978-1-4666-3649-1.ch003

Griswold, W. (2013). Transformative learning and educational technology integration in a post-totalitarian context: Professional development among school teachers in rural Siberia, Russia. In E. Jean Francois (Ed.), *Transcultural blended learning and teaching in postsecondary education* (pp. 128–144). Hershey, PA: IGI Global. doi:10.4018/978-1-4666-2014-8.ch008

Gruich, M. R. (2013). Defining professional development for technology. In S. Wang & T. Hartsell (Eds.), *Technology integration and foundations for effective leadership* (pp. 152–170). Hershey, PA: IGI Global. doi:10.4018/978-1-4666-2656-0.ch009

Guidry, K. R., & Pasquini, L. (2013). Twitter chat as an informal learning tool: A case study using #sachat. In H. Yang & S. Wang (Eds.), *Cases on formal and informal e-learning environments: Opportunities and practices* (pp. 356–377). Hershey, PA: IGI Global. doi:10.4018/978-1-4666-1930-2.ch019

Gunn, C., & Blake, A. (2009). Blending technology into an academic practice qualification for university teachers. In E. Stacey & P. Gerbic (Eds.), *Effective blended learning practices: Evidence-based perspectives in ICT-facilitated education* (pp. 259–279). Hershey, PA: IGI Global. doi:10.4018/978-1-60566-296-1.ch014

Hanewald, R. (2013). Professional development with and for emerging technologies: A case study with Asian languages and cultural studies teachers in Australia. In J. Keengwe (Ed.), *Pedagogical applications and social effects of mobile technology integration* (pp. 175–192). Hershey, PA: IGI Global. doi:10.4018/978-1-4666-2985-1.ch010

Hansen, C. C. (2012). ABCs and PCs: Effective professional development in early childhood education. In I. Chen & D. McPheeters (Eds.), *Cases on educational technology integration in urban schools* (pp. 230–235). Hershey, PA: IGI Global. doi:10.4018/978-1-61350-492-5.ch032

Hao, S. (2012). Turn on your mobile devices: Potential and considerations of informal mobile learning. In V. Dennen & J. Myers (Eds.), *Virtual professional development and informal learning via social networks* (pp. 39–58). Hershey, PA: IGI Global. doi:10.4018/978-1-4666-1815-2.ch003

Harteis, C. (2010). Contributions of e-collaborative knowledge construction to professional learning and expertise. In B. Ertl (Ed.), *E-collaborative knowledge construction: Learning from computer-supported and virtual environments* (pp. 91–108). Hershey, PA: IGI Global. doi:10.4018/978-1-61520-729-9.ch005

Hartsell, T., Herron, S. S., Fang, H., & Rathod, A. (2011). Improving teachers' self-confidence in learning technology skills and math education through professional development. In *Instructional design: Concepts, methodologies, tools and applications* (pp. 1487–1503). Hershey, PA: IGI Global. doi:10.4018/978-1-60960-503-2.ch609

Hartsell, T., Herron, S. S., Fang, H., & Rathod, A. (2012). Improving teachers' self-confidence in learning technology skills and math education through professional development. In L. Tomei (Ed.), *Advancing education with information communication technologies: Facilitating new trends* (pp. 150–164). Hershey, PA: IGI Global. doi:10.4018/978-1-61350-468-0.ch013

Hartsell, T., & Wang, S. (2013). Introduction to technology integration and leadership. In S. Wang & T. Hartsell (Eds.), *Technology integration and foundations for effective leadership* (pp. 1–17). Hershey, PA: IGI Global. doi:10.4018/978-1-4666-2656-0.ch001

Hauge, T. E., & Norenes, S. O. (2010). VideoPaper as a bridging tool in teacher professional development. In J. Lindberg & A. Olofsson (Eds.), *Online learning communities and teacher professional development: Methods for improved education delivery* (pp. 209–228). Hershey, PA: IGI Global. doi:10.4018/978-1-60566-780-5.ch012

Haythornthwaite, C., & De Laat, M. (2012). Social network informed design for learning with educational technology. In A. Olofsson & J. Lindberg (Eds.), *Informed design of educational technologies in higher education: Enhanced learning and teaching* (pp. 352–374). Hershey, PA: IGI Global. doi:10.4018/978-1-61350-080-4.ch018

Heinrichs, L., Fellander-Tsai, L., & Davies, D. (2013). Clinical virtual worlds: The wider implications for professional development in healthcare. In K. Bredl & W. Bösche (Eds.), *Serious games and virtual worlds in education, professional development, and healthcare* (pp. 221–240). Hershey, PA: IGI Global. doi:10.4018/978-1-4666-3673-6.ch014

Helleve, I. (2010). Theoretical foundations of teachers' professional development. In J. Lindberg & A. Olofsson (Eds.), *Online learning communities and teacher professional development: Methods for improved education delivery* (pp. 1–19). Hershey, PA: IGI Global. doi:10.4018/978-1-60566-780-5.ch001

Hemphill, L. S., & McCaw, D. S. (2009). Moodling professional development training that worked. In L. Tan Wee Hin & R. Subramaniam (Eds.), *Handbook of research on new media literacy at the K-12 level: Issues and challenges* (pp. 808–822). Hershey, PA: IGI Global. doi:10.4018/978-1-60566-120-9.ch050

Hensley, M. K. (2010). Teaching new librarians how to teach: A model for building a peer learning program. In E. Pankl, D. Theiss-White, & M. Bushing (Eds.), *Recruitment, development, and retention of information professionals: Trends in human resources and knowledge management* (pp. 179–190). Hershey, PA: IGI Global. doi:10.4018/978-1-61520-601-8.ch010

Hernández-Gantes, V. M. (2011). Helping faculty design online courses in higher education. In V. Wang (Ed.), *Encyclopedia of information communication technologies and adult education integration* (pp. 779–794). Hershey, PA: IGI Global. doi:10.4018/978-1-61692-906-0.ch047

Herrington, J., & Oliver, R. (2006). Professional development for the online teacher: An authentic approach. In T. Herrington & J. Herrington (Eds.), *Authentic learning environments in higher education* (pp. 283–295). Hershey, PA: IGI Global. doi:10.4018/978-1-59140-594-8.ch020

Hicks, T. (2013). Adding the "digital layer": Examining one teacher's growth as a digital writer through an NWP summer institute and beyond. In K. Pytash & R. Ferdig (Eds.), *Exploring technology for writing and writing instruction* (pp. 345–357). Hershey, PA: IGI Global. doi:10.4018/978-1-4666-4341-3.ch020

Hinson, J. M., & Bordelon Sellers, R. (2004). Professional development recommendations for online courses. In C. Cavanaugh (Ed.), *Development and management of virtual schools: Issues and trends* (pp. 135–157). Hershey, PA: IGI Global. doi:10.4018/978-1-59140-154-4.ch007

Hirtle, J., & Smith, S. (2010). When virtual communities click: Transforming teacher practice, transforming teachers. In H. Yang & S. Yuen (Eds.), *Handbook of research on practices and outcomes in e-learning: Issues and trends* (pp. 182–196). Hershey, PA: IGI Global. doi:10.4018/978-1-60566-788-1.ch011

Holland, I. E. (2007). Evolution of the Milwaukee public schools portal. In A. Tatnall (Ed.), *Encyclopedia of portal technologies and applications* (pp. 397–401). Hershey, PA: IGI Global. doi:10.4018/978-1-59140-989-2.ch067

Holt, J., Unruh, L., & Dougherty, A. M. (2011). Enhancing a rural school-university teacher education partnership through an e-mentoring program for beginning teachers. In M. Bowdon & R. Carpenter (Eds.), *Higher education, emerging technologies, and community partnerships: Concepts, models and practices* (pp. 212–220). Hershey, PA: IGI Global. doi:10.4018/978-1-60960-623-7.ch019

Hood, D. W., & Huang, W. D. (2013). Professional development with graduate teaching assistants (TAs) teaching online. In J. Keengwe & L. Kyei-Blankson (Eds.), *Virtual mentoring for teachers: Online professional development practices* (pp. 26–42). Hershey, PA: IGI Global. doi:10.4018/978-1-4666-1963-0.ch002

Hu, X. C., & Meyen, E. L. (2013). A comparison of student and instructor preferences for design and pedagogy features in postsecondary online courses. In M. Raisinghani (Ed.), *Curriculum, learning, and teaching advancements in online education* (pp. 213–229). Hershey, PA: IGI Global. doi:10.4018/978-1-4666-2949-3.ch015

Hucks, D., & Ragan, M. (2013). Technology expanding horizons in teacher education: Transformative learning experiences. In J. Keengwe (Ed.), *Research perspectives and best practices in educational technology integration* (pp. 61–79). Hershey, PA: IGI Global. doi:10.4018/978-1-4666-2988-2.ch004

Hui, D., & Russell, D. L. (2009). Understanding the effectiveness of collaborative activity in online professional development with innovative educators through intersubjectivity. In L. Tomei (Ed.), *Information communication technologies for enhanced education and learning: Advanced applications and developments* (pp. 283–302). Hershey, PA: IGI Global. doi:10.4018/978-1-60566-150-6.ch020

Hulme, M., & Hughes, J. (2006). Patchwork e-dialogues in the professional development of new teachers. In J. O'Donoghue (Ed.), *Technology supported learning and teaching: A staff perspective* (pp. 210–223). Hershey, PA: IGI Global. doi:10.4018/978-1-59140-962-5.ch013

Hunt, L., & Sankey, M. (2013). Getting the context right for quality teaching and learning. In D. Salter (Ed.), *Cases on quality teaching practices in higher education* (pp. 261–279). Hershey, PA: IGI Global. doi:10.4018/978-1-4666-3661-3.ch016

Hur, J. W., Brush, T., & Bonk, C. (2012). An analysis of teacher knowledge and emotional sharing in a teacher blog community. In V. Dennen & J. Myers (Eds.), *Virtual professional development and informal learning via social networks* (pp. 219–239). Hershey, PA: IGI Global. doi:10.4018/978-1-4666-1815-2.ch013

Hurst, A. (2012). Reflections on personal experiences of staff training and continuing professional development for academic staff in the development of high quality support for disabled students in higher education. In D. Moore, A. Gorra, M. Adams, J. Reaney, & H. Smith (Eds.), *Disabled students in education: Technology, transition, and inclusivity* (pp. 288–304). Hershey, PA: IGI Global. doi:10.4018/978-1-61350-183-2.ch015

Hyatt, K. J. (2011). Technology: A reflective tool for professional development. In L. Tomei (Ed.), *Online courses and ICT in education: Emerging practices and applications* (pp. 134–142). Hershey, PA: IGI Global. doi:10.4018/978-1-60960-150-8.ch011

Jackson, T. (2012). Ways to mentor methods' faculty integration of technologies in their courses. In D. Polly, C. Mims, & K. Persichitte (Eds.), *Developing technology-rich teacher education programs: Key issues* (pp. 519–534). Hershey, PA: IGI Global. doi:10.4018/978-1-4666-0014-0.ch033

Jamieson-Proctor, R., & Finger, G. (2009). Measuring and evaluating ICT use: Developing an instrument for measuring student ICT use. In L. Tan Wee Hin & R. Subramaniam (Eds.), *Handbook of research on new media literacy at the K-12 level: Issues and challenges* (pp. 326–339). Hershey, PA: IGI Global. doi:10.4018/978-1-60566-120-9.ch021

Jarvis, D. H. (2012). Teaching mathematics teachers online: Strategies for navigating the intersection of andragogy, technology, and reform-based mathematics education. In A. Juan, M. Huertas, S. Trenholm, & C. Steegmann (Eds.), *Teaching mathematics online: Emergent technologies and methodologies* (pp. 187–199). Hershey, PA: IGI Global. doi:10.4018/978-1-60960-875-0.ch009

Jimoyiannis, A., Gravani, M., & Karagiorgi, Y. (2012). Teacher professional development through virtual campuses: Conceptions of a 'new' model. In H. Yang & S. Yuen (Eds.), *Handbook of research on practices and outcomes in virtual worlds and environments* (pp. 327–347). Hershey, PA: IGI Global. doi:10.4018/978-1-60960-762-3.ch017

Johnson, E. S., & Pitcock, J. (2008). Preparing online instructors: Beyond using the technology. In K. Orvis & A. Lassiter (Eds.), *Computer-supported collaborative learning: Best practices and principles for instructors* (pp. 89–113). Hershey, PA: IGI Global. doi:10.4018/978-1-59904-753-9.ch005

Johnson, K., & Tashiro, J. (2010). Interprofessional care and health care complexity: Factors shaping human resources effectiveness in health information management. In S. Kabene (Ed.), *Human resources in healthcare, health informatics and healthcare systems* (pp. 250–280). Hershey, PA: IGI Global. doi:10.4018/978-1-61520-885-2.ch015

Johnson, K., & Tashiro, J. (2013). Interprofessional care and health care complexity: Factors shaping human resources effectiveness in health information management. In *User-driven healthcare: Concepts, methodologies, tools, and applications* (pp. 1273–1302). Hershey, PA: IGI Global. doi:10.4018/978-1-4666-2770-3.ch064

Johnson, M. L. (2014). Moving from theory to practice: Integrating personal learning networks into a graduate-level student development theory course. In S. Leone (Ed.), *Synergic integration of formal and informal e-learning environments for adult lifelong learners* (pp. 165–177). Hershey, PA: IGI Global. doi:10.4018/978-1-4666-4655-1.ch008

Johnson, V. (2009). Understanding dynamic change and creation of learning organizations. In P. Rogers, G. Berg, J. Boettcher, C. Howard, L. Justice, & K. Schenk (Eds.), *Encyclopedia of distance learning* (2nd ed.; pp. 2187–2191). Hershey, PA: IGI Global. doi:10.4018/978-1-60566-198-8.ch323

Jones, M. G., & Harris, L. (2012). Using student choice to promote technology integration: The buffet model. In D. Polly, C. Mims, & K. Persichitte (Eds.), *Developing technology-rich teacher education programs: Key issues* (pp. 192–204). Hershey, PA: IGI Global. doi:10.4018/978-1-4666-0014-0.ch013

Joyes, G., Fisher, T., Firth, R., & Coyle, D. (2014). The nature of a successful online professional doctorate. In K. Sullivan, P. Czigler, & J. Sullivan Hellgren (Eds.), *Cases on professional distance education degree programs and practices: Successes, challenges, and issues* (pp. 296–330). Hershey, PA: IGI Global. doi:10.4018/978-1-4666-4486-1.ch011

Kelly, D. (2009). Modeling best practices in web-based academic development. In R. Donnelly & F. McSweeney (Eds.), *Applied e-learning and e-teaching in higher education* (pp. 35–55). Hershey, PA: IGI Global. doi:10.4018/978-1-59904-814-7.ch003

Kelly, D. K. (2010). Modeling best practices in web-based academic development. In A. Tatnall (Ed.), *Web technologies: Concepts, methodologies, tools, and applications* (pp. 1578–1595). Hershey, PA: IGI Global. doi:10.4018/978-1-60566-982-3.ch085

Kelly, R. (2014). Administration: Making a connection with the library's strongest advocate. In K. Kennedy & L. Green (Eds.), *Collaborative models for librarian and teacher partnerships* (pp. 175–184). Hershey, PA: IGI Global. doi:10.4018/978-1-4666-4361-1.ch015

Kennedy-Clark, S., & Thompson, K. (2013). A MUVEing success: Design strategies for professional development in the use of multi-user virtual environments and educational games in science education. In S. D'Agustino (Ed.), *Immersive environments, augmented realities, and virtual worlds: Assessing future trends in education* (pp. 16–41). Hershey, PA: IGI Global. doi:10.4018/978-1-4666-2670-6.ch002

Kent, A. M. (2013). Teacher leadership: Learning and leading. In J. Lewis, A. Green, & D. Surry (Eds.), *Technology as a tool for diversity leadership: Implementation and future implications* (pp. 230–242). Hershey, PA: IGI Global. doi:10.4018/978-1-4666-2668-3.ch018

Keppell, M. J. (2007). Instructional designers on the borderline: Brokering across communities of practice. In M. Keppell (Ed.), *Instructional design: Case studies in communities of practice* (pp. 68–89). Hershey, PA: IGI Global. doi:10.4018/978-1 59904-322-7.ch004

Kidd, T. T., & Keengwe, J. (2012). Technology integration and urban schools: Implications for instructional practices. In L. Tomei (Ed.), *Advancing education with information communication technologies: Facilitating new trends* (pp. 244–256). Hershey, PA: IGI Global. doi:10.4018/978-1-61350-468-0.ch020

King, K. P. (2008). The transformation model. In C. Van Slyke (Ed.), *Information communication technologies: Concepts, methodologies, tools, and applications* (pp. 1102–1108). Hershey, PA: IGI Global. doi:10.4018/978-1-59904-949-6.ch073

King, K. P. (2012). Impact of podcasts as professional learning: Teacher created, student created, and professional development podcasts. In S. Chhabra (Ed.), *ICTs for advancing rural communities and human development: Addressing the digital divide* (pp. 237–250). Hershey, PA: IGI Global. doi:10.4018/978-1-4666-0047-8.ch016

Kitchenham, A. (2011). Blending professional development for rural educators an exploratory study. In D. Parsons (Ed.), *Combining e-learning and m-learning: New applications of blended educational resources* (pp. 225–238). Hershey, PA: IGI Global. doi:10.4018/978-1-60960-481-3.ch014

Klieger, A., & Oster-Levinz, A. (2010). How online tasks promote teachers' expertise within the technological pedagogical content knowledge (TPACK). In T. Yuzer & G. Kurubacak (Eds.), *Transformative learning and online education: Aesthetics, dimensions and concepts* (pp. 219–235). Hershey, PA: IGI Global. doi:10.4018/978-1-61520-985-9.ch015

Koh, S., Lee, S., Yen, D. C., & Havelka, D. (2005). Information technology professional career development: A progression of skills. In M. Hunter & F. Tan (Eds.), *Advanced topics in global information management* (Vol. 4, pp. 142–157). Hershey, PA: IGI Global. doi:10.4018/978-1-59140-468-2.ch009

Kopcha, T., & Valentine, K. D. (2013). A framework for developing robust online professional development materials to support teacher practice under the common core. In D. Polly (Ed.), *Common core mathematics standards and implementing digital technologies* (pp. 319–331). Hershey, PA: IGI Global. doi:10.4018/978-1-4666-4086-3.ch021

Koumpis, A. (2010). Culture of services. In A. Koumpis (Ed.), *Service science for socio-economical and information systems advancement: Holistic methodologies* (pp. 312–347). Hershey, PA: IGI Global. doi:10.4018/978-1-60566-683-9.ch010

Kyei-Blankson, L., & Keengwe, J. (2013). Faculty-faculty interactions in online learning environments. In L. Tomei (Ed.), *Learning tools and teaching approaches through ICT advancements* (pp. 127–135). Hershey, PA: IGI Global. doi:10.4018/978-1-4666-2017-9.ch012

Larson, L., & Vanmetre, S. (2010). Learning together with the interactive white board. In N. Lambropoulos & M. Romero (Eds.), *Educational social software for context-aware learning: Collaborative methods and human interaction* (pp. 69–78). Hershey, PA: IGI Global. doi:10.4018/978-1-60566-826-0.ch005

Laurillard, D., & Masterman, E. (2010). TPD as online collaborative learning for innovation in teaching. In J. Lindberg & A. Olofsson (Eds.), *Online learning communities and teacher professional development: Methods for improved education delivery* (pp. 230–246). Hershey, PA: IGI Global. doi:10.4018/978-1-60566-780-5.ch013

Lawrence, J., Burton, L., Summers, J., Noble, K., & Gibbings, P. D. (2013). An associate dean's community of practice: Rising to the leadership challenges of engaging distance students using blended models of learning and teaching. In J. Willems, B. Tynan, & R. James (Eds.), *Global challenges and perspectives in blended and distance learning* (pp. 212–222). Hershey, PA: IGI Global. doi:10.4018/978-1-4666-3978-2.ch017

Lee, D. M. (2004). Organizational entry and transition from academic study: Examining a critical step in the professional development of young IS workers. In M. Igbaria & C. Shayo (Eds.), *Strategies for managing IS/IT personnel* (pp. 113–142). Hershey, PA: IGI Global. doi:10.4018/978-1-59140-128-5.ch006

Leh, A. S., & Grafton, L. (2009). Promoting new media literacy in a school district. In L. Tan Wee Hin & R. Subramaniam (Eds.), *Handbook of research on new media literacy at the K-12 level: Issues and challenges* (pp. 607–619). Hershey, PA: IGI Global. doi:10.4018/978-1-60566-120-9.ch038

Lehew, A. J., & Polly, D. (2013). The use of digital resources to support elementary school teachers' implementation of the common core state standards. In D. Polly (Ed.), *Common core mathematics standards and implementing digital technologies* (pp. 332–338). Hershey, PA: IGI Global. doi:10.4018/978-1-4666-4086-3.ch022

Leng, J., & Sharrock, W. (2010). Collaborative practices in computer-aided academic research. In I. Portela & M. Cruz-Cunha (Eds.), *Information communication technology law, protection and access rights: Global approaches and issues* (pp. 249–270). Hershey, PA: IGI Global. doi:10.4018/978-1-61520-975-0.ch016

Ley, K., & Gannon-Cook, R. (2010). Marketing a blended university program: An action research case study. In S. Mukerji & P. Tripathi (Eds.), *Cases on technology enhanced learning through collaborative opportunities* (pp. 73–90). Hershey, PA: IGI Global. doi:10.4018/978-1-61520-751-0.ch005

Linton, J., & Stegall, D. (2013). Common core standards for mathematical practice and TPACK: An integrated approach to instruction. In D. Polly (Ed.), *Common core mathematics standards and implementing digital technologies* (pp. 234–249). Hershey, PA: IGI Global. doi:10.4018/978-1-4666-4086-3.ch016

Lithgow, C. M., Wolf, J. L., & Berge, Z. L. (2011). Virtual worlds: Corporate early adopters pave the way. In S. Hai-Jew (Ed.), *Virtual immersive and 3D learning spaces: Emerging technologies and trends* (pp. 25–43). Hershey, PA: IGI Global. doi:10.4018/978-1-61692-825-4.ch002

Lloyd, M., & Duncan-Howell, J. (2010). Changing the metaphor: The potential of online communities in teacher professional development. In J. Lindberg & A. Olofsson (Eds.), *Online learning communities and teacher professional development: Methods for improved education delivery* (pp. 60–76). Hershey, PA: IGI Global. doi:10.4018/978-1-60566-780-5.ch004

Lockyer, L., Patterson, J., Rowland, G., & Hearne, D. (2007). ActiveHealth: Enhancing the community of physical and health educators through online technologies. In M. Keppell (Ed.), *Instructional design: Case studies in communities of practice* (pp. 331–348). Hershey, PA: IGI Global. doi:10.4018/978-1-59904-322-7.ch017

Loose, C. (2013). Teachers as researchers and instructional leaders. In V. Wang (Ed.), *Handbook of research on teaching and learning in K-20 education* (pp. 710–725). Hershey, PA: IGI Global. doi:10.4018/978-1-4666-4249-2.ch041

Luetkehans, L. M., & Hunt, R. D. (2014). School librarians as significant other: Using online professional learning communities for the development of pre-service teachers. In K. Kennedy & L. Green (Eds.), *Collaborative models for librarian and teacher partnerships* (pp. 56–66). Hershey, PA: IGI Global. doi:10.4018/978-1-4666-4361-1.ch006

Lyublinskaya, I., & Tournaki, N. (2012). The effects of teacher content authoring on TPACK and on student achievement in algebra: Research on instruction with the TI-Nspire™ handheld. In R. Ronau, C. Rakes, & M. Niess (Eds.), *Educational technology, teacher knowledge, and classroom impact: A research handbook on frameworks and approaches* (pp. 295–322). Hershey, PA: IGI Global. doi:10.4018/978-1-60960-750-0.ch013

Mackey, J. (2009). Virtual learning and real communities: Online professional development for teachers. In E. Stacey & P. Gerbic (Eds.), *Effective blended learning practices: Evidence-based perspectives in ICT-facilitated education* (pp. 163–181). Hershey, PA: IGI Global. doi:10.4018/978-1-60566-296-1.ch009

Mackey, J., & Mills, A. (2003). An examination of ICT planning maturity in schools: A stage theory perspective. In T. McGill (Ed.), *Current issues in IT education* (pp. 376–395). Hershey, PA: IGI Global. doi:10.4018/978-1-93177-753-7.ch030

Manathunga, C., & Donnelly, R. (2009). Opening online academic development programmes to international perspectives and dialogue. In R. Donnelly & F. McSweeney (Eds.), *Applied e-learning and e-teaching in higher education* (pp. 85–109). Hershey, PA: IGI Global. doi:10.4018/978-1-59904-814-7.ch005

Marinho, R. (2010). Faculty development in instructional technology in the context of learning styles and institutional barriers. In S. Mukerji & P. Tripathi (Eds.), *Cases on interactive technology environments and transnational collaboration: Concerns and perspectives* (pp. 1–38). Hershey, PA: IGI Global. doi:10.4018/978-1-61520-909-5.ch001

Mark, C. L. (2013). Evaluating and funding the professional development program. In S. Wang & T. Hartsell (Eds.), *Technology integration and foundations for effective leadership* (pp. 206–226). Hershey, PA: IGI Global. doi:10.4018/978-1-4666-2656-0.ch012

Marshall, J. C. (2013). Measuring and facilitating highly effective inquiry-based teaching and learning in science classrooms. In M. Khine & I. Saleh (Eds.), *Approaches and strategies in next generation science learning* (pp. 290–306). Hershey, PA: IGI Global. doi:10.4018/978-1-4666-2809-0.ch015

Marshall, K. (2008). E-portfolios in teacher education. In G. Putnik & M. Cruz-Cunha (Eds.), *Encyclopedia of networked and virtual organizations* (pp. 516–523). Hershey, PA: IGI Global. doi:10.4018/978-1-59904-885-7.ch068

Martin, M. (2008). Integrating videoconferencing into the classroom: A perspective from Northern Ireland. In D. Newman, J. Falco, S. Silverman, & P. Barbanell (Eds.), *Videoconferencing technology in K-12 instruction: Best practices and trends* (pp. 253–268). Hershey, PA: IGI Global. doi:10.4018/978-1-59904-331-9.ch019

Maurer, M. J. (2012). Telementoring and virtual professional development: A theoretical perspective from science on the roles of self-efficacy, teacher learning, and professional learning communities. In *Organizational learning and knowledge: Concepts, methodologies, tools and applications* (pp. 1158–1176). Hershey, PA: IGI Global. doi:10.4018/978-1-60960-783-8.ch319

McAnuff-Gumbs, M., & Verbeck, K. (2013). Toward a model of multi-level professional learning communities to guide the training and practice of literacy coaches. In H. Yang & S. Wang (Eds.), *Cases on online learning communities and beyond: Investigations and applications* (pp. 361–402). Hershey, PA: IGI Global. doi:10.4018/978-1-4666-1936-4.ch019

McCarthy, J. (2012). Connected: Online mentoring in Facebook for final year digital media students. In A. Okada, T. Connolly, & P. Scott (Eds.), *Collaborative learning 2.0: Open educational resources* (pp. 204–221). Hershey, PA: IGI Global. doi:10.4018/978-1-4666-0300-4.ch011

McConnell, D. (2005). Networked collaborative e-learning. In E. Li & T. Du (Eds.), *Advances in electronic business* (Vol. 1, pp. 222–257). Hershey, PA: IGI Global. doi:10.4018/978-1-59140-381-4.ch008

McCormack, V. (2010). Utilizing VoiceThread to increase teacher candidates' reflection and global implications for usage. In J. Yamamoto, J. Kush, R. Lombard, & C. Hertzog (Eds.), *Technology implementation and teacher education: Reflective models* (pp. 108–123). Hershey, PA: IGI Global. doi:10.4018/978-1-61520-897-5.ch007

McGrath, E., Lowes, S., McKay, M., Sayres, J., & Lin, P. (2012). Robots underwater! Learning science, engineering and 21st century skills: The evolution of curricula, professional development and research in formal and informal contexts. In B. Barker, G. Nugent, N. Grandgenett, & V. Adamchuk (Eds.), *Robots in K-12 education: A new technology for learning* (pp. 141–167). Hershey, PA: IGI Global. doi:10.4018/978-1-4666-0182-6.ch007

McGuigan, A. (2012). Blogospheric learning in a continuing professional development context. In A. Okada, T. Connolly, & P. Scott (Eds.), *Collaborative learning 2.0: Open educational resources* (pp. 222–237). Hershey, PA: IGI Global. doi:10.4018/978-1-4666-0300-4.ch012

McIntosh, S. (2005). Expanding the classroom: Using online discussion forums in college and professional development courses. In K. St.Amant & P. Zemliansky (Eds.), *Internet-based workplace communications: Industry and academic applications* (pp. 68–87). Hershey, PA: IGI Global. doi:10.4018/978-1-59140-521-4.ch004

McNair, V., & Marshall, K. (2006). How eportfolios support development in early teacher education. In A. Jafari & C. Kaufman (Eds.), *Handbook of research on eportfolios* (pp. 474–485). Hershey, PA: IGI Global. doi:10.4018/978-1-59140-890-1.ch042

McNair, V., & Marshall, K. (2008). How eportfolios support development in early teacher education. In L. Tomei (Ed.), *Online and distance learning: Concepts, methodologies, tools, and applications* (pp. 2130–2137). Hershey, PA: IGI Global. doi:10.4018/978-1-59904-935-9.ch172

McPherson, M., Baptista Nunes, M., Sandars, J., & Kell, C. (2008). Technology and continuing professional education: The reality beyond the hype. In T. Kidd & I. Chen (Eds.), *Social information technology: Connecting society and cultural issues* (pp. 296–312). Hershey, PA: IGI Global. doi:10.4018/978-1-59904-774-4.ch019

Medina, A. L., Tobin, M. T., Pilonieta, P., Chiappone, L. L., & Blanton, W. E. (2012). Application of computer, digital, and telecommunications technologies to the clinical preparation of teachers. In D. Polly, C. Mims, & K. Persichitte (Eds.), *Developing technology-rich teacher education programs: Key issues* (pp. 480–498). Hershey, PA: IGI Global. doi:10.4018/978-1-4666-0014-0.ch031

Meletiou-Mavrotheris, M. (2012). Online communities of practice as vehicles for teacher professional development. In A. Juan, M. Huertas, S. Trenholm, & C. Steegmann (Eds.), *Teaching mathematics online: Emergent technologies and methodologies* (pp. 142–166). Hershey, PA: IGI Global. doi:10.4018/978-1-60960-875-0.ch007

Meltzer, S. T. (2013). The impact of new technologies on professional development. In V. Bryan & V. Wang (Eds.), *Technology use and research approaches for community education and professional development* (pp. 40–52). Hershey, PA: IGI Global. doi:10.4018/978-1-4666-2955-4.ch003

Milman, N. B., Hillarious, M., O'Neill, V., & Walker, B. (2013). Going 1:1 with laptop computers in an independent, co-educational middle and high school. In J. Keengwe (Ed.), *Pedagogical applications and social effects of mobile technology integration* (pp. 156–174). Hershey, PA: IGI Global. doi:10.4018/978-1-4666-2985-1.ch009

Moon, B. (2012). Teaching teachers: The biggest educational challenge in sub-Saharan Africa. In R. Hogan (Ed.), *Transnational distance learning and building new markets for universities* (pp. 198–209). Hershey, PA: IGI Global. doi:10.4018/978-1-4666-0206-9.ch012

Mørch, A. I., & Andersen, R. (2012). Mutual development: The software engineering context of end-user development. In A. Dwivedi & S. Clarke (Eds.), *End-user computing, development, and software engineering: New challenges* (pp. 103–125). Hershey, PA: IGI Global. doi:10.4018/978-1-4666-0140-6.ch005

Morrow, D., & Bagnall, R. G. (2010). Hybridizing online learning with external interactivity. In F. Wang, J. Fong, & R. Kwan (Eds.), *Handbook of research on hybrid learning models: Advanced tools, technologies, and applications* (pp. 24–41). Hershey, PA: IGI Global. doi:10.4018/978-1-60566-380-7.ch002

Mountain, L. A. (2008). Videoconferencing: An alternative to traditional professional development in the K-12 setting. In D. Newman, J. Falco, S. Silverman, & P. Barbanell (Eds.), *Videoconferencing technology in K-12 instruction: Best practices and trends* (pp. 213–225). Hershey, PA: IGI Global. doi:10.4018/978-1-59904-331-9.ch016

Mouzakis, C., & Bourletidis, C. (2010). A blended learning course for teachers' ongoing professional development in Greece. In J. Yamamoto, J. Kush, R. Lombard, & C. Hertzog (Eds.), *Technology implementation and teacher education: Reflective models* (pp. 1–24). Hershey, PA: IGI Global. doi:10.4018/978-1-61520-897-5.ch001

Mouzakis, C., Tsaknakis, H., & Tziortzioti, C. (2012). Theoretical rationale for designing a blended learning teachers' professional development program. In P. Anastasiades (Ed.), *Blended learning environments for adults: Evaluations and frameworks* (pp. 274–289). Hershey, PA: IGI Global. doi:10.4018/978-1-4666-0939-6.ch014

Murphy, M. G., & Calway, P. B. (2011). Continuing professional development: Work and learning integration for professionals. In P. Keleher, A. Patil, & R. Harreveld (Eds.), *Work-integrated learning in engineering, built environment and technology: Diversity of practice in practice* (pp. 25–51). Hershey, PA: IGI Global. doi:10.4018/978-1-60960-547-6.ch002

Mustapha, W. Z. (2012). The art and science of designing and developing an online English language training module for adult learners. In N. Alias & S. Hashim (Eds.), *Instructional technology research, design and development: Lessons from the field* (pp. 270–286). Hershey, PA: IGI Global. doi:10.4018/978-1-61350-198-6.ch017

Mutohar, A., & Hughes, J. E. (2013). Toward web 2.0 integration in indonesian education: Challenges and planning strategies. In N. Azab (Ed.), *Cases on web 2.0 in developing countries: Studies on implementation, application, and use* (pp. 198–221). Hershey, PA: IGI Global. doi:10.4018/978-1-4666-2515-0.ch008

Newman, D. L., Clure, G., Deyoe, M. M., & Connor, K. A. (2013). Using technology in a studio approach to learning: Results of a five year study of an innovative mobile teaching tool. In J. Keengwe (Ed.), *Pedagogical applications and social effects of mobile technology integration* (pp. 114–132). Hershey, PA: IGI Global. doi:10.4018/978-1-4666-2985-1.ch007

Newman, D. L., Coyle, V. C., & McKenna, L. A. (2013). Changing the face of ELA classrooms: A case study of TPACK professional development. In J. Keengwe (Ed.), *Research perspectives and best practices in educational technology integration* (pp. 270–287). Hershey, PA: IGI Global. doi:10.4018/978-1-4666-2988-2.ch014

Ng, A. W., & Ho, F. (2014). Dynamics of knowledge renewal for professional accountancy under globalization. In P. Ordóñez de Pablos & R. Tennyson (Eds.), *Strategic approaches for human capital management and development in a turbulent economy* (pp. 264–278). Hershey, PA: IGI Global. doi:10.4018/978-1-4666-4530-1.ch017

Ng, F. F. (2005). Knowledge management in higher education and professional development in the construction industry. In A. Kazi (Ed.), *Knowledge management in the construction industry: A socio-technical perspective* (pp. 150–165). Hershey, PA: IGI Global. doi:10.4018/978-1-59140-360-9.ch009

Ng, F. F. (2008). Knowledge management in higher education and professional development in the construction industry. In M. Jennex (Ed.), *Knowledge management: Concepts, methodologies, tools, and applications* (pp. 2355–2368). Hershey, PA: IGI Global. doi:10.4018/978-1-59904-933-5.ch195

Nguyen, V., & Szymanski, M. (2013). A state of the art cart: Visual arts and technology integration in teacher education. In J. Keengwe (Ed.), *Research perspectives and best practices in educational technology integration* (pp. 80–104). Hershey, PA: IGI Global. doi:10.4018/978-1-4666-2988-2.ch005

Niemitz, M., Slough, S., St. John, K., Leckie, R. M., Peart, L., & Klaus, A. (2010). Integrating K-12 hybrid online learning activities in teacher education programs: Reflections from the school of rock expedition. In J. Yamamoto, J. Kush, R. Lombard, & C. Hertzog (Eds.), *Technology implementation and teacher education: Reflective models* (pp. 25–43). Hershey, PA: IGI Global. doi:10.4018/978-1-61520-897-5.ch002

Norris, D. M. (2005). Driving systemic change with e-learning. In C. Howard, J. Boettcher, L. Justice, K. Schenk, P. Rogers, & G. Berg (Eds.), *Encyclopedia of distance learning* (pp. 687–695). Hershey, PA: IGI Global. doi:10.4018/978-1-59140-555-9.ch100

Northrup, P. T., & Harrison, W. T. Jr. (2011). Using learning objects for rapid deployment to mobile learning devices for the U.S. coast guard. In *Instructional design: Concepts, methodologies, tools and applications* (pp. 527–540). Hershey, PA: IGI Global. doi:10.4018/978-1-60960-503-2.ch303

Northrup, P. T., Rasmussen, K. L., & Dawson, D. B. (2004). Designing and reusing learning objects to streamline WBI development. In A. Armstrong (Ed.), *Instructional design in the real world: A view from the trenches* (pp. 184–200). Hershey, PA: IGI Global. doi:10.4018/978-1-59140-150-6.ch011

Northrup, P. T., Rasmussen, K. L., & Dawson, D. B. (2008). Designing and reusing learning objects to streamline WBI development. In S. Clarke (Ed.), *End-user computing: Concepts, methodologies, tools, and applications* (pp. 1–1). Hershey, PA: IGI Global. doi:10.4018/978-1-59904-945-8.ch037

Oigara, J. N. (2013). Integrating technology in teacher education programs. In J. Keengwe (Ed.), *Research perspectives and best practices in educational technology integration* (pp. 28–43). Hershey, PA: IGI Global. doi:10.4018/978-1-4666-2988-2.ch002

Orrill, C. H., & Polly, D. (2012). Technology integration in mathematics: A model for integrating technology through content development. In D. Polly, C. Mims, & K. Persichitte (Eds.), *Developing technology-rich teacher education programs: Key issues* (pp. 337–356). Hershey, PA: IGI Global. doi:10.4018/978-1-4666-0014-0.ch022

Ostashewski, N., & Reid, D. (2013). The networked learning framework: A model for networked professional learning utilizing social networking sites. In J. Keengwe & L. Kyei-Blankson (Eds.), *Virtual mentoring for teachers: Online professional development practices* (pp. 66–83). Hershey, PA: IGI Global. doi:10.4018/978-1-4666-1963-0.ch004

Ostashewski, N., & Reid, D. (2013). The iPad in the classroom: Three implementation cases highlighting pedagogical activities, integration issues, and teacher professional development strategies. In J. Keengwe (Ed.), *Pedagogical applications and social effects of mobile technology integration* (pp. 25–41). Hershey, PA: IGI Global. doi:10.4018/978-1-4666-2985-1.ch002

Pachler, N., Daly, C., & Turvey, A. (2010). Teacher professional development practices: The case of the haringey transformation teachers programme. In J. Lindberg & A. Olofsson (Eds.), *Online learning communities and teacher professional development: Methods for improved education delivery* (pp. 77–95). Hershey, PA: IGI Global. doi:10.4018/978-1-60566-780-5.ch005

Parker, D. (2013). Implementing the professional development program. In S. Wang & T. Hartsell (Eds.), *Technology integration and foundations for effective leadership* (pp. 190–205). Hershey, PA: IGI Global. doi:10.4018/978-1-4666-2656-0.ch011

Payne, A. (2013). Designing a professional development program. In S. Wang & T. Hartsell (Eds.), *Technology integration and foundations for effective leadership* (pp. 171–189). Hershey, PA: IGI Global. doi:10.4018/978-1-4666-2656-0.ch010

Peacock, S., & Dunlop, G. M. (2006). Developing e-learning provision for healthcare professionals' continuing professional development. In J. O'Donoghue (Ed.), *Technology supported learning and teaching: A staff perspective* (pp. 106–124). Hershey, PA: IGI Global. doi:10.4018/978-1-59140-962-5.ch007

Piecka, D. C., Ruberg, L., Ruckman, C., & Fullwood, D. (2012). NASATalk as a discovery learning space: Self-discovery learning opportunities. In S. Hai-Jew (Ed.), *Constructing self-discovery learning spaces online: Scaffolding and decision making technologies* (pp. 49–71). Hershey, PA: IGI Global. doi:10.4018/978-1-61350-320-1.ch004

Pilkington, R. (2010). Building practitioner skills in personalised elearning: Messages for professional development. In J. O'Donoghue (Ed.), *Technology-supported environments for personalized learning: Methods and case studies* (pp. 167–184). Hershey, PA: IGI Global. doi:10.4018/978-1-60566-884-0.ch010

Polly, D. (2011). Preparing teachers to integrate technology effectively: The case of higher-order thinking skills (HOTS). In S. D'Agustino (Ed.), *Adaptation, resistance and access to instructional technologies: Assessing future trends in education* (pp. 395–409). Hershey, PA: IGI Global. doi:10.4018/978-1-61692-854-4.ch023

Polly, D. (2013). Designing and teaching an online elementary mathematics methods course: Promises, barriers, and implications. In R. Hartshorne, T. Heafner, & T. Petty (Eds.), *Teacher education programs and online learning tools: Innovations in teacher preparation* (pp. 335–356). Hershey, PA: IGI Global. doi:10.4018/978-1-4666-1906-7.ch018

Polly, D., Grant, M. M., & Gikas, J. (2011). Supporting technology integration in higher education: The role of professional development. In D. Surry, R. Gray Jr, & J. Stefurak (Eds.), *Technology integration in higher education: Social and organizational aspects* (pp. 58–71). Hershey, PA: IGI Global. doi:10.4018/978-1-60960-147-8.ch005

Polly, D., Mims, C., & McCombs, B. (2012). Designing district-wide technology-rich professional development. In I. Chen & D. McPheeters (Eds.), *Cases on educational technology integration in urban schools* (pp. 236–243). Hershey, PA: IGI Global. doi:10.4018/978-1-61350-492-5.ch033

Powell, E. (2009). Facilitating reflective teaching: Video-stimulated reflective dialogues as a professional development process. In V. Wang (Ed.), *Handbook of research on e-learning applications for career and technical education: Technologies for vocational training* (pp. 100–111). Hershey, PA: IGI Global. doi:10.4018/978-1-60566-739-3.ch008

Prisk, J., & Lee, K. (2012). How to utilize an online community of practice (CoP) to enhance innovation in teaching and learning. In V. Wang (Ed.), *Encyclopedia of e-leadership, counseling and training* (pp. 532–544). Hershey, PA: IGI Global. doi:10.4018/978-1-61350-068-2.ch039

Prpic, J. K., & Moore, G. (2012). E-portfolios as a quantitative and qualitative means of demonstrating learning outcomes and competencies in engineering. In K. Yusof, N. Azli, A. Kosnin, S. Yusof, & Y. Yusof (Eds.), *Outcome-based science, technology, engineering, and mathematics education: Innovative practices* (pp. 124–154). Hershey, PA: IGI Global. doi:10.4018/978-1-4666-1809-1.ch007

Pullman, N., & Streff, K. (2009). Creating a security education, training, and awareness program. In M. Gupta & R. Sharman (Eds.), *Handbook of research on social and organizational liabilities in information security* (pp. 325–345). Hershey, PA: IGI Global. doi:10.4018/978-1-60566-132-2.ch020

Quinton, S. (2007). Delivering online expertise, online. In M. Keppell (Ed.), *Instructional design: Case studies in communities of practice* (pp. 193–214). Hershey, PA: IGI Global. doi:10.4018/978-1-59904-322-7.ch010

Reali, A. M., Tancredi, R. M., & Mizukami, M. D. (2012). Online mentoring as a tool for professional development and change of novice and experienced teachers: A Brazilian experience. In V. Dennen & J. Myers (Eds.), *Virtual professional development and informal learning via social networks* (pp. 203–218). Hershey, PA: IGI Global. doi:10.4018/978-1-4666-1815-2.ch012

Redmon, R. J. Jr. (2009). E-mail reflection groups as collaborative action research. In J. Salmons & L. Wilson (Eds.), *Handbook of research on electronic collaboration and organizational synergy* (pp. 349–361). Hershey, PA: IGI Global. doi:10.4018/978-1-60566-106-3.ch023

Rice, M. L., & Bain, C. (2013). Planning and implementation of a 21st century classroom project. In A. Benson, J. Moore, & S. Williams van Rooij (Eds.), *Cases on educational technology planning, design, and implementation: A project management perspective* (pp. 76–92). Hershey, PA: IGI Global. doi:10.4018/978-1-4666-4237-9.ch005

Richardson, S. L., Barnes, S. L., & Torain, D. S. (2012). Using technology to support algebra teaching and assessment: A teacher development case study. In I. Chen & D. McPheeters (Eds.), *Cases on educational technology integration in urban schools* (pp. 224–229). Hershey, PA: IGI Global. doi:10.4018/978-1-61350-492-5.ch031

Rieber, L. P., Francom, G. M., & Jensen, L. J. (2011). Feeling like a first year teacher: Toward becoming a successful online instructor. In D. Surry, R. Gray Jr, & J. Stefurak (Eds.), *Technology integration in higher education: Social and organizational aspects* (pp. 42–57). Hershey, PA: IGI Global. doi:10.4018/978-1-60960-147-8.ch004

Ring, G., & Foti, S. (2006). Using eportfolios to facilitate professional development among pre-service teachers. In A. Jafari & C. Kaufman (Eds.), *Handbook of research on eportfolios* (pp. 340–357). Hershey, PA: IGI Global. doi:10.4018/978-1-59140-890-1.ch031

Riverin, S. (2009). Blended learning and professional development in the K-12 sector. In E. Stacey & P. Gerbic (Eds.), *Effective blended learning practices: Evidence-based perspectives in ICT-facilitated education* (pp. 182–202). Hershey, PA: IGI Global. doi:10.4018/978-1-60566-296-1.ch010

Robertshaw, M. B., Leary, H., Walker, A., Bloxham, K., & Recker, M. (2009). Reciprocal mentoring "in the wild": A retrospective, comparative case study of ICT teacher professional development. In E. Stacey & P. Gerbic (Eds.), *Effective blended learning practices: evidence-based perspectives in ICT-facilitated education* (pp. 280–297). Hershey, PA: IGI Global. doi:10.4018/978-1-60566-296-1.ch015

Robertson, L., & Hardman, W. (2013). More than changing classrooms: Professors' transitions to synchronous e-teaching. In P. Ordóñez de Pablos & R. Tennyson (Eds.), *Strategic role of tertiary education and technologies for sustainable competitive advantage* (pp. 156–175). Hershey, PA: IGI Global. doi:10.4018/978-1-4666-4233-1.ch007

Rockland, R., Kimmel, H., Carpinelli, J., Hirsch, L. S., & Burr-Alexander, L. (2012). Medical robotics in K-12 education. In B. Barker, G. Nugent, N. Grandgenett, & V. Adamchuk (Eds.), *Robots in K-12 education: A new technology for learning* (pp. 120–140). Hershey, PA: IGI Global. doi:10.4018/978-1-4666-0182-6.ch006

Rockland, R., Kimmel, H., Carpinelli, J., Hirsch, L. S., & Burr-Alexander, L. (2014). Medical robotics in K-12 education. In Robotics: Concepts, methodologies, tools, and applications (pp. 1096-1115). Hershey, PA: IGI Global. doi:10.4018/978-1-4666-4607-0.ch053

Rodesiler, L., & Tripp, L. (2012). It's all about personal connections: Pre-service English teachers' experiences engaging in networked learning. In V. Dennen & J. Myers (Eds.), *Virtual professional development and informal learning via social networks* (pp. 185–202). Hershey, PA: IGI Global. doi:10.4018/978-1-4666-1815-2.ch011

Ronau, R. N., & Rakes, C. R. (2012). A comprehensive framework for teacher knowledge (CFTK): Complexity of individual aspects and their interactions. In R. Ronau, C. Rakes, & M. Niess (Eds.), *Educational technology, teacher knowledge, and classroom impact: A research handbook on frameworks and approaches* (pp. 59–102). Hershey, PA: IGI Global. doi:10.4018/978-1-60960-750-0.ch004

Rosen, Y., & Rimor, R. (2013). Teaching and assessing problem solving in online collaborative environment. In R. Hartshorne, T. Heafner, & T. Petty (Eds.), *Teacher education programs and online learning tools: Innovations in teacher preparation* (pp. 82–97). Hershey, PA: IGI Global. doi:10.4018/978-1-4666-1906-7.ch005

Ruberg, L., Calinger, M., & Howard, B. C. (2010). Evaluating educational technologies: Historical milestones. In L. Tomei (Ed.), *ICTs for modern educational and instructional advancement: New approaches to teaching* (pp. 285–297). Hershey, PA: IGI Global. doi:10.4018/978-1-60566-936-6.ch024

Russell, D. L. (2007). The mediated action of educational reform: An inquiry into collaboative online professional development. In R. Sharma & S. Mishra (Eds.), *Cases on global e-learning practices: Successes and pitfalls* (pp. 108–122). Hershey, PA: IGI Global. doi:10.4018/978-1-59904-340-1.ch009

Sáenz, J., Aramburu, N., & Rivera, O. (2010). Exploring the links between structural capital, knowledge sharing, innovation capability and business competitiveness: An empirical study. In D. Harorimana (Ed.), *Cultural implications of knowledge sharing, management and transfer: Identifying competitive advantage* (pp. 321–354). Hershey, PA: IGI Global. doi:10.4018/978-1-60566-790-4.ch015

Sales, G. C. (2009). Preparing teachers to teach online. In P. Rogers, G. Berg, J. Boettcher, C. Howard, L. Justice, & K. Schenk (Eds.), *Encyclopedia of distance learning* (2nd ed.; pp. 1665–1672). Hershey, PA: IGI Global. doi:10.4018/978-1-60566-198-8.ch244

Sales, G. C. (2011). Preparing teachers to teach online. In *Instructional design: Concepts, methodologies, tools and applications* (pp. 8–17). Hershey, PA: IGI Global. doi:10.4018/978-1-60960-503-2.ch102

Sampson, D. G., & Kallonis, P. (2012). 3D virtual classroom simulations for supporting school teachers' continuing professional development. In J. Jia (Ed.), *Educational stages and interactive learning: From kindergarten to workplace training* (pp. 427–450). Hershey, PA: IGI Global. doi:10.4018/978-1-4666-0137-6.ch023

Sari, E., & Lim, C. P. (2012). Online learning community: building the professional capacity of Indonesian teachers. In J. Jia (Ed.), *Educational stages and interactive learning: From kindergarten to workplace training* (pp. 451–467). Hershey, PA: IGI Global. doi:10.4018/978-1-4666-0137-6.ch024

Scheckler, R. (2010). Case studies from the inquiry learning forum: Stories reaching beyond the edges. In J. Lindberg & A. Olofsson (Eds.), *Online learning communities and teacher professional development: Methods for improved education delivery* (pp. 42–59). Hershey, PA: IGI Global. doi:10.4018/978-1-60566-780-5.ch003

Schifter, C. (2008). "Making teachers better": A brief history of professional development for teachers. In C. Schifter (Ed.), *Infusing technology into the classroom: Continuous practice improvement* (pp. 41–57). Hershey, PA: IGI Global. doi:10.4018/978-1-59904-765-2.ch003

Schifter, C. (2008). Effecting change in the classroom through professional development. In C. Schifter (Ed.), *Infusing technology into the classroom: Continuous practice improvement* (pp. 259–274). Hershey, PA: IGI Global. doi:10.4018/978-1-59904-765-2.ch014

Schifter, C. (2008). Continuous practice improvement. In C. Schifter (Ed.), *Infusing technology into the classroom: Continuous practice improvement* (pp. 58–86). Hershey, PA: IGI Global. doi:10.4018/978-1-59904-765-2.ch004

Schifter, C. (2008). Finger painting to digital painting: First grade. In C. Schifter (Ed.), *Infusing technology into the classroom: Continuous practice improvement* (pp. 109–126). Hershey, PA: IGI Global. doi:10.4018/978-1-59904-765-2.ch006

Schrader, P., Strudler, N., Asay, L., Graves, T., Pennell, S. L., & Stewart, S. (2012). The pathway to Nevada's future: A case of statewide technology integration and professional development. In I. Chen & D. McPheeters (Eds.), *Cases on educational technology integration in urban schools* (pp. 204–223). Hershey, PA: IGI Global. doi:10.4018/978-1-61350-492-5.ch030

Scott, D. E., & Scott, S. (2010). Innovations in the use of technology and teacher professional development. In J. Lindberg & A. Olofsson (Eds.), *Online learning communities and teacher professional development: Methods for improved education delivery* (pp. 169–189). Hershey, PA: IGI Global. doi:10.4018/978-1-60566-780-5.ch010

Scott, D. E., & Scott, S. (2012). Multi-faceted professional development models designed to enhance teaching and learning within universities. In A. Olofsson & J. Lindberg (Eds.), *Informed design of educational technologies in higher education: Enhanced learning and teaching* (pp. 412–435). Hershey, PA: IGI Global. doi:10.4018/978-1-61350-080-4.ch021

Scott, S. (2010). The theory and practice divide in relation to teacher professional development. In J. Lindberg & A. Olofsson (Eds.), *Online learning communities and teacher professional development: Methods for improved education delivery* (pp. 20–40). Hershey, PA: IGI Global. doi:10.4018/978-1-60566-780-5.ch002

Semich, G. W., & Gibbons, B. (2011). The professional development school: A building block for training public school faculty on new technologies. In L. Tomei (Ed.), *Online courses and ICT in education: Emerging practices and applications* (pp. 99–108). Hershey, PA: IGI Global. doi:10.4018/978-1-60960-150-8.ch008

Shambaugh, N. (2013). A professional development school technology integration and research plan. In A. Ritzhaupt & S. Kumar (Eds.), *Cases on educational technology implementation for facilitating learning* (pp. 45–68). Hershey, PA: IGI Global. doi:10.4018/978-1-4666-3676-7.ch003

Sherman, G., & Byers, A. (2011). Electronic portfolios in the professional development of educators. In S. D'Agustino (Ed.), *Adaptation, resistance and access to instructional technologies: Assessing future trends in education* (pp. 429–444). Hershey, PA: IGI Global. doi:10.4018/978-1-61692-854-4.ch025

Simelane, S. (2010). Professional development programme in the use of educational technology to implement technology-enhanced courses successfully. In S. Mukerji & P. Tripathi (Eds.), *Cases on technology enhanced learning through collaborative opportunities* (pp. 91–110). Hershey, PA: IGI Global. doi:10.4018/978-1-61520-751-0.ch006

Skibba, K. (2013). Adult learning influence on faculty learning cycle: Individual and shared reflections while learning to teach online lead to pedagogical transformations. In J. Keengwe & L. Kyei-Blankson (Eds.), *Virtual mentoring for teachers: Online professional development practices* (pp. 263–291). Hershey, PA: IGI Global. doi:10.4018/978-1-4666-1963-0.ch014

Skinner, L. B., Witte, M. M., & Witte, J. E. (2010). Challenges and opportunities in career and technical education. In V. Wang (Ed.), *Definitive readings in the history, philosophy, theories and practice of career and technical education* (pp. 197–215). Hershey, PA: IGI Global. doi:10.4018/978-1-61520-747-3.ch012

Slabon, W. A., & Richards, R. L. (2012). Story-based professional development: Using a conflict management wiki. In V. Dennen & J. Myers (Eds.), *Virtual professional development and informal learning via social networks* (pp. 256–275). Hershey, PA: IGI Global. doi:10.4018/978-1-4666-1815-2.ch015

Spaulding, D. T. (2008). Virtual field trips: Advantages and disadvantages for educators and recommendation for professional development. In D. Newman, J. Falco, S. Silverman, & P. Barbanell (Eds.), *Videoconferencing technology in K-12 instruction: Best practices and trends* (pp. 191–199). Hershey, PA: IGI Global. doi:10.4018/978-1-59904-331-9.ch014

Speaker, R. B., Levitt, G., & Grubaugh, S. (2013). Professional development in a virtual world. In J. Keengwe & L. Kyei-Blankson (Eds.), *Virtual mentoring for teachers: Online professional development practices* (pp. 122–148). Hershey, PA: IGI Global. doi:10.4018/978-1-4666-1963-0.ch007

Stacey, E., & Gerbic, P. (2009). Introduction to blended learning practices. In E. Stacey & P. Gerbic (Eds.), *Effective blended learning practices: Evidence-based perspectives in ICT-facilitated education* (pp. 1–19). Hershey, PA: IGI Global. doi:10.4018/978-1-60566-296-1.ch001

Stanfill, D. (2012). Standards-based educational technology professional development. In V. Wang (Ed.), *Encyclopedia of e-leadership, counseling and training* (pp. 819–834). Hershey, PA: IGI Global. doi:10.4018/978-1-61350-068-2.ch060

Steel, C., & Andrews, T. (2012). Re-imagining teaching for technology-enriched learning spaces: An academic development model. In M. Keppell, K. Souter, & M. Riddle (Eds.), *Physical and virtual learning spaces in higher education: Concepts for the modern learning environment* (pp. 242–265). Hershey, PA: IGI Global. doi:10.4018/978-1-60960-114-0.ch015

Stewart, C., Horarik, S., & Wolodko, K. (2013). Maximising technology usage in research synthesis of higher education professional development research. In J. Willems, B. Tynan, & R. James (Eds.), *Global challenges and perspectives in blended and distance learning* (pp. 1–16). Hershey, PA: IGI Global. doi:10.4018/978-1-4666-3978-2.ch001

Stieha, V., & Raider-Roth, M. (2012). Disrupting relationships: A catalyst for growth. In J. Faulkner (Ed.), *Disrupting pedagogies in the knowledge society: Countering conservative norms with creative approaches* (pp. 16–31). Hershey, PA: IGI Global. doi:10.4018/978-1-61350-495-6.ch002

Stockero, S. L. (2010). Serving rural teachers using synchronous online professional development. In J. Yamamoto, J. Leight, S. Winterton, & C. Penny (Eds.), *Technology leadership in teacher education: Integrated solutions and experiences* (pp. 111–124). Hershey, PA: IGI Global. doi:10.4018/978-1-61520-899-9.ch007

Stylianou-Georgiou, A., Vrasidas, C., Christodoulou, N., Zembylas, M., & Landone, E. (2006). Technologies challenging literacy: Hypertext, community building, reflection, and critical literacy. In L. Tan Wee Hin & R. Subramaniam (Eds.), *Handbook of research on literacy in technology at the K-12 level* (pp. 21–33). Hershey, PA: IGI Global. doi:10.4018/978-1-59140-494-1.ch002

Szecsy, E. M. (2011). Building knowledge without borders: Using ICT to develop a binational education research community. In G. Kurubacak & T. Yuzer (Eds.), *Handbook of research on transformative online education and liberation: Models for social equality* (pp. 67–85). Hershey, PA: IGI Global. doi:10.4018/978-1-60960-046-4.ch004

Tawfik, A. A., Reiseck, C., & Richter, R. (2013). Project management methods for the implementation of an online faculty development course. In A. Benson, J. Moore, & S. Williams van Rooij (Eds.), *Cases on educational technology planning, design, and implementation: A project management perspective* (pp. 153–167). Hershey, PA: IGI Global. doi:10.4018/978-1-4666-4237-9.ch009

Taylor, D. B., Hartshorne, R., Eneman, S., Wilkins, P., & Polly, D. (2012). Lessons learned from the implementation of a technology-focused professional learning community. In D. Polly, C. Mims, & K. Persichitte (Eds.), *Developing technology-rich teacher education programs: Key issues* (pp. 535–550). Hershey, PA: IGI Global. doi:10.4018/978-1-4666-0014-0.ch034

Tedford, D. (2010). Perspectives on the influences of social capital upon internet usage of rural Guatemalan teachers. In S. Mukerji & P. Tripathi (Eds.), *Cases on technological adaptability and transnational learning: Issues and challenges* (pp. 218–243). Hershey, PA: IGI Global. doi:10.4018/978-1-61520-779-4.ch012

Terantino, J. M. (2012). An activity theoretical approach to examining virtual professional development and informal learning via social networks. In V. Dennen & J. Myers (Eds.), *Virtual professional development and informal learning via social networks* (pp. 60–74). Hershey, PA: IGI Global. doi:10.4018/978-1-4666-1815-2.ch004

Thompson, T. L., & Kanuka, H. (2009). Establishing communities of practice for effective and sustainable professional development for blended learning. In E. Stacey & P. Gerbic (Eds.), *Effective blended learning practices: Evidence-based perspectives in ICT-facilitated education* (pp. 144–162). Hershey, PA: IGI Global. doi:10.4018/978-1-60566-296-1.ch008

Thornton, J. (2010). Framing pedagogy, diminishing technology: Teachers experience of online learning software. In H. Song & T. Kidd (Eds.), *Handbook of research on human performance and instructional technology* (pp. 263–283). Hershey, PA: IGI Global. doi:10.4018/978-1-60566-782-9.ch016

Thornton, K., & Yoong, P. (2010). The application of blended action learning to leadership development: A case study. In P. Yoong (Ed.), *Leadership in the digital enterprise: Issues and challenges* (pp. 163–180). Hershey, PA: IGI Global. doi:10.4018/978-1-60566-958-8.ch012

Ting, A., & Jones, P. D. (2010). Using free source eportfolios to empower ESL teachers in collaborative peer reflection. In J. Yamamoto, J. Kush, R. Lombard, & C. Hertzog (Eds.), *Technology implementation and teacher education: Reflective models* (pp. 93–107). Hershey, PA: IGI Global. doi:10.4018/978-1-61520-897-5.ch006

Tomei, L. A. (2008). The KARPE model revisited – An updated investigation for differentiating teaching and learning with technology in higher education. In L. Tomei (Ed.), *Adapting information and communication technologies for effective education* (pp. 30–40). Hershey, PA: IGI Global. doi:10.4018/978-1-59904-922-9.ch003

Torrisi-Steele, G. (2005). Toward effective use of multimedia technologies in education. In S. Mishra & R. Sharma (Eds.), *Interactive multimedia in education and training* (pp. 25–46). Hershey, PA: IGI Global. doi:10.4018/978-1-59140-393-7.ch002

Torrisi-Steele, G. (2008). Toward effective use of multimedia technologies in education. In M. Syed (Ed.), *Multimedia technologies: Concepts, methodologies, tools, and applications* (pp. 1651–1667). Hershey, PA: IGI Global. doi:10.4018/978-1-59904-953-3.ch118

Trujillo, K. M., Wiburg, K., Savic, M., & McKee, K. (2013). Teachers learn how to effectively integrate mobile technology by teaching students using math snacks animations and games. In J. Keengwe (Ed.), *Pedagogical applications and social effects of mobile technology integration* (pp. 98–113). Hershey, PA: IGI Global. doi:10.4018/978-1-4666-2985-1.ch006

Tynan, B., & Barnes, C. (2010). Web 2.0 and professional development of academic staff. In M. Lee & C. McLoughlin (Eds.), *Web 2.0-based e-learning: Applying social informatics for tertiary teaching* (pp. 365–379). Hershey, PA: IGI Global. doi:10.4018/978-1-60566-294-7.ch019

Tynan, B., & Barnes, C. (2012). Web 2.0 and professional development of academic staff. In *Virtual learning environments: Concepts, methodologies, tools and applications* (pp. 94–108). Hershey, PA: IGI Global. doi:10.4018/978-1-4666-0011-9.ch107

Uehara, D. L. (2007). Research in the Pacific: Utilizing technology to inform and improve teacher practice. In Y. Inoue (Ed.), *Technology and diversity in higher education: New challenges* (pp. 213–232). Hershey, PA: IGI Global. doi:10.4018/978-1-59904-316-6.ch011

Velez-Solic, A., & Banas, J. R. (2013). Professional development for online educators: Problems, predictions, and best practices. In J. Keengwe & L. Kyei-Blankson (Eds.), *Virtual mentoring for teachers: Online professional development practices* (pp. 204–226). Hershey, PA: IGI Global. doi:10.4018/978-1-4666-1963-0.ch011

Venkatesh, V., Bures, E., Davidson, A., Wade, C. A., Lysenko, L., & Abrami, P. C. (2013). Electronic portfolio encouraging active and reflective learning: A case study in improving academic self-regulation through innovative use of educational technologies. In A. Ritzhaupt & S. Kumar (Eds.), *Cases on educational technology implementation for facilitating learning* (pp. 341–376). Hershey, PA: IGI Global. doi:10.4018/978-1-4666-3676-7.ch019

Watson, C. E., Zaldivar, M., & Summers, T. (2010). ePortfolios for learning, assessment, and professional development. In R. Donnelly, J. Harvey, & K. O'Rourke (Eds.), Critical design and effective tools for e-learning in higher education: Theory into practice (pp. 157-175). Hershey, PA: IGI Global. doi:10.4018/978-1-61520-879-1.ch010

Whitehouse, P., McCloskey, E., & Ketelhut, D. J. (2010). Online pedagogy design and development: New models for 21st century online teacher professional development. In J. Lindberg & A. Olofsson (Eds.), *Online learning communities and teacher professional development: Methods for improved education delivery* (pp. 247–262). Hershey, PA: IGI Global. doi:10.4018/978-1-60566-780-5.ch014

Williams, I. M., & Olaniran, B. A. (2012). Professional development through web 2.0 collaborative applications. In V. Dennen & J. Myers (Eds.), *Virtual professional development and informal learning via social networks* (pp. 1–24). Hershey, PA: IGI Global. doi:10.4018/978-1-4666-1815-2.ch001

Wilson, A., & Christie, D. (2010). Realising the potential of virtual environments: A challenge for Scottish teachers. In J. Lindberg & A. Olofsson (Eds.), *Online learning communities and teacher professional development: Methods for improved education delivery* (pp. 96–113). Hershey, PA: IGI Global. doi:10.4018/978-1-60566-780-5.ch006

Wilson, G. (2009). Case studies of ICT-enhanced blended learning and implications for professional development. In E. Stacey & P. Gerbic (Eds.), *Effective blended learning practices: Evidence-based perspectives in ICT-facilitated education* (pp. 239–258). Hershey, PA: IGI Global. doi:10.4018/978-1-60566-296-1.ch013

Witt, L. A., & Burke, L. A. (2003). Using cognitive ability and personality to select information technology professionals. In M. Mahmood (Ed.), *Advanced topics in end user computing* (Vol. 2, pp. 1–17). Hershey, PA: IGI Global. doi:10.4018/978-1-59140-065-3.ch001

Wynne, C. W. (2014). Cultivating leaders from within: Transforming workers into leaders. In S. Mukerji & P. Tripathi (Eds.), *Handbook of research on transnational higher education* (pp. 42–58). Hershey, PA: IGI Global. doi:10.4018/978-1-4666-4458-8.ch003

Yakavenka, H. (2012). Developing professional competencies through international peer learning communities. In V. Dennen & J. Myers (Eds.), *Virtual professional development and informal learning via social networks* (pp. 134–154). Hershey, PA: IGI Global. doi:10.4018/978-1-4666-1815-2.ch008

Yamamoto, J., Linaberger, M., & Forbes, L. S. (2005). Mentoring and technology integration for teachers. In D. Carbonara (Ed.), *Technology literacy applications in learning environments* (pp. 161–170). Hershey, PA: IGI Global. doi:10.4018/978-1-59140-479-8.ch012

Yukawa, J. (2011). Telementoring and project-based learning: An integrated model for 21st century skills. In D. Scigliano (Ed.), *Telementoring in the K-12 classroom: Online communication technologies for learning* (pp. 31–56). Hershey, PA: IGI Global. doi:10.4018/978-1-61520-861-6.ch003

Zellermayer, M., Mor, N., & Heilweil, I. (2009). The intersection of theory, tools and tasks in a postgraduate learning environment. In C. Payne (Ed.), *Information technology and constructivism in higher education: Progressive learning frameworks* (pp. 319–333). Hershey, PA: IGI Global. doi:10.4018/978-1-60566-654-9.ch021

Zuidema, L. A. (2008). Parawork. In P. Zemliansky & K. St.Amant (Eds.), *Handbook of research on virtual workplaces and the new nature of business practices* (pp. 81–97). Hershey, PA: IGI Global. doi:10.4018/978-1-59904-893-2.ch007

Zygouris-Coe, V. I. (2013). A model for online instructor training, support, and professional development. In J. Keengwe & L. Kyei-Blankson (Eds.), *Virtual mentoring for teachers: Online professional development practices* (pp. 97–121). Hershey, PA: IGI Global. doi:10.4018/978-1-4666-1963-0.ch006

Zygouris-Coe, V. I., & Swan, B. (2010). Challenges of online teacher professional development communities: A statewide case study in the United States. In J. Lindberg & A. Olofsson (Eds.), *Online learning communities and teacher professional development: Methods for improved education delivery* (pp. 114–133). Hershey, PA: IGI Global. doi:10.4018/978-1-60566-780-5.ch007

About the Contributors

Adedeji Tella, Ph.D., is a lecturer in the department of science and Technology Education, Faculty of Education, University of Ibadan, Nigeria. He holds a Ph.D in Teacher Education with focus in Mathematics Education and Curriculum Instruction from University of Ibadan. In addition he holds M.Ed in Mathematics Education and Curriculum Instruction, B.Ed in Mathematics/Chemistry from the same university and Nigerian Certificate in Education (NCE) in Mathematics/Chemistry from Osun State College of Education Ilesa. His research focuses on Mathematics and sciences instructional effectiveness and quality improvement, ICT in mathematics and Science Education in learning mathematics, mathematics curriculum design and misconception. Psychological variables as predictors of learning mathematics/ sciences and research methodologies. Dr Tella has published research activities in the above areas at both local and international reputable refereed journals. He has presented papers at various international conferences, contributed chapters to a number of books and encyclopedias, and co-authored monetho. Thirty mathematics textbooks and e-books for both primary and secondary schools in Nigeria for reputable publishers e.g Evans Publisher Nigeria Ltd, Nelson Publisher Nigeria Ltd and others. He is a seasoned public examinations examiners and accessor to Port Forte Harvey University, South Africa for Postgraduate students. He is also an accessor to many learned journals and member of learned professional associations. He is currently the Subdean Postgraduate, Faculty of Education, University of Ibadan from 1 Aug. 2017 – 31 July 2019. His current research focuses on Left handedness and students' learning outcomes in mathematics.

* * *

Adebowale Adebagbo is a well Trained Teacher and certified educational technologist who is passionate about making quality teaching and learning to be easy, flexible, enjoyable and productive at all levels. Adebowale is self-motivated on solutions to solve societal problems. He has produced a number of local board games to teach mathematics skills and some topics in other subjects. He always engages strongly in innovation, big or small, to do things in a better way. He has contributed well researched articles and book chapter to academic publications, locally and internationally. He has more than two decades of teaching experience across Secondary schools, Colleges of Education and Universities. His academic qualifications are - N.C.E (T) in Building Construction Technology Education, from Federal College of Education (Technical), Akoka, Lagos; B.Ed in Educational Technology from University of Ilorin, Nigeria; M.Ed and Ph.D in Educational Technology, both from University of Ibadan, Nigeria. Currently, he is lecturing degree students of University of Nigeria, Nsukka, at Yaba College of Technology, Nigeria.

Saheed Ayodeji Adejimi is a doctoral student of the African Centre of Excellence for Innovative Teaching and Learning Mathematics and Science, University of Rwanda, College of Education Rukara Campus.

Yetunde O. Akanle was born in Ibadan, Oyo state, Nigeria in 1976.She is a passionate reader and writer. She had her first degree in Pure and Applied Mathematics (M.Tech), from Ladoke Akintola University of Technology. Masters in Mathematics (M.Sc) and Mathematics Education (M.Ed) from University of Ibadan. She is a lecturer in one of the colleges of Agriculture in Nigeria. She is happily married with children.

Faith Chidinma Nworah is from Anambra State Nigeria, was born in Sokoto State Nigeria, studied Education Mathematics at Usmanu Danfodio University Sokoto, currently running her M.Ed at the University of Ibadan also a member of Mathematical Association of Nigeria (MAN).

Efe O. Ogbebor received her M.Ed and B.Ed degrees in chemistry education and curriculum studies from University of Ibadan, Nigeria. Her research focus is on science education, innovative and active strategies of chemistry and science teaching. Currently she is working on her Ph.D.

Ayodeji Ojo was born in Akure, Ondo State. He received his B.Sc.Ed with Honors in Biology from Adekunle Ajasin University, Akungba Akoko in Ondo State and his M.Ed in Science Education from University of Ibadan. His primary research interests are in the field of Biology Education pedagogy and theory. Specifically, he is interested in student active learning, and pedagogies of experimental learning. Currently he is a Ph.D student at the University of Ibadan.

Index

P

S

T

www.ingramcontent.com/pod-product-compliance
Lightning Source LLC
LaVergne TN
LVHW061534290926
849162LV00004B/792